Howard

Ocean Life

Ocean Life

Time-Life Books   Alexandria, Virginia

Table of Contents

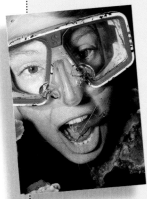

Ocean Life

Look out over the blue, rolling waters of the ocean and what do you see? What appears to be an endless, almost empty body of water. But dive down into it and you'll discover a world teeming with life. Within the ocean are countless **habitats** filled with plants and **animals** specially adapted to live there. Most **species** live in the shallows and near the surface where there is abundant sunlight to support plant life. But even the dark, frigid ocean depths harbor life forms —including some very strange ones!

Flip through the pages of this book and you'll meet some amazing creatures, including glow-in-the-dark fish, a squid as long as two school buses, and animals so tiny that 1,000 of them could fit on the head of a pin. Dive in and discover the wonders of our watery planet.

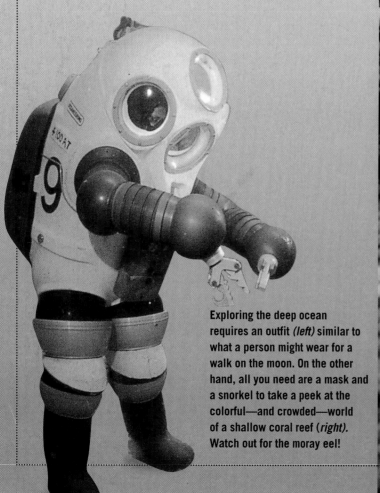

Exploring the deep ocean requires an outfit *(left)* similar to what a person might wear for a walk on the moon. On the other hand, all you need are a mask and a snorkel to take a peek at the colorful—and crowded—world of a shallow coral reef *(right)*. Watch out for the moray eel!

Our Watery World

Looking back at the distant earth from the blackness of space, one of the Apollo astronauts thought our home resembled "a big blue marble." That's not surprising since water covers two-thirds of the planet's surface. The earth is the only planet in the solar system known to contain liquid water. And it was in this ocean water that simple life forms first grew more than three billion years ago.

The earth's massive ocean is one continuous body of water linked around the globe. But to make identification easier, geographers have divided it into four separate oceans: the Pacific, the Atlantic, the Indian, and the Arctic. Sometimes the water surrounding Antarctica is referred to as the Southern Ocean. All together, these oceans contain more than 1.35 billion cubic km (324 million cubic mi.) of water.

On a map, the earth's continents appear to float like islands in a vast sea of blue.

Fast FACTS

Pacific Ocean The largest ocean covers 181,340,000 sq km (70,015,000 sq mi.), or nearly one-third of the earth's surface.

Atlantic Ocean The second largest ocean covers 94,310,000 sq km (36,413,000 sq mi.), or about one-fifth of the earth.

Indian Ocean Nearly as large as the Atlantic, this ocean covers 74,120,000 sq km (28,618,000 sq mi.).

Arctic Ocean The smallest and shallowest ocean at 12,260,000 sq km (4,734,000 sq mi.), it is always almost entirely covered with ice about 3 m (10 ft.) thick.

Mariana Trench The deepest point on the earth's surface—11,035 m (36,205 ft.)—was caused by the collision of two **tectonic plates** under the Pacific Ocean.

Mid-Ocean Ridge The world's longest mountain range—65,000 km (40,000 mi.)—snakes over one-third of the ocean floor.

How the Oceans Formed

Above and below the Ocean Waves

Stingrays take it easy in the azure blue waters of the Caribbean Sea around the Cayman Islands. In the sky above, clouds form from evaporated seawater, powerful evidence of the ocean's role in making weather (pages 22-23).

What's an Oceanographer?

A scuba-diving oceanographer records field notes on a clipboard using a special waterproof pen. Oceanographers, also known as **marine** scientists, study all aspects of the world's oceans: the seawater itself and the terrain of the ocean floor, currents, and tides; the effects of pollution; and the plants and animals living in the oceans.

What Are the Seven Seas?

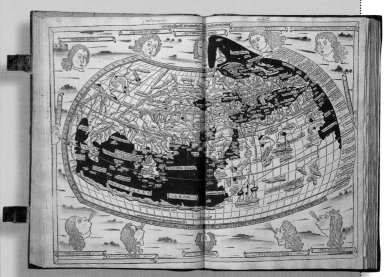

Four and a half billion years ago, the earth was a very young planet. It had no liquid water. But there was a lot of water **vapor,** which is water in **gaseous** form, trapped within the molten rock under the earth's crust. This water vapor was pushed to the surface and released into the air through the many volcanic eruptions that shook the land.

After about a half billion years, the earth's crust began to cool. Water vapor in the atmosphere **condensed,** forming huge storm clouds. Torrents of rain fell continuously for tens of millions of years, creating enormous, raging rivers that flooded vast low-lying areas. Icy comets collided with the new planet, adding more water to the growing seas. By studying the oldest known rocks, scientists have determined that oceans first appeared on earth about 3.8 billion years ago.

The main difference between an ocean and a sea is size. Oceans are much bigger and deeper, whereas seas are partially or almost completely enclosed by land. This map in an early atlas, published in Germany in 1482, shows the known world before explorers circled the globe and proved how large the oceans really are. People at that time thought there were many seas surrounded by a lot of land. People often used the word "seven" to mean "many." So the Seven Seas became a common name for all the world's large bodies of water.

Ocean Floor

Until this century, it was thought that the ocean floor was simply a thick, flat layer of mud. We now know it is a landscape of extremes— deep canyons, enormous mountains, and vast plains—as shown in the illustration at right.

From the land's edge, the shallow, gently sloping continental shelf teems with **marine** life. Its floor holds valuable resources such as oil and natural gas. The steep continental slope drops off quickly to the abyssal plain, the flattest place on earth, which stretches to the mid-ocean ridge. Narrow valleys, or **trenches,** and soaring volcanic seamounts dot the ocean floor. Flat-topped seamounts are called guyots.

In the center of every ocean is a winding chain of mountains called the mid-ocean ridge, a feature that occurs where **tectonic plates** are being pushed apart by the upward flow of molten rock. As the rock cools, the ocean floor grows larger by about 1 cm (0.4 in.) a year.

Sonar

A sonar scan *(right)* displays a 30-km (19-mi.)-wide section of Australia's Great Barrier Reef. To create the picture, oceanographers towed a torpedo-shaped sonar instrument below their ship. Sonar works by sending pulses of sound energy downward and sideways. A computer records how long it takes the sound waves to bounce off the ocean floor and return to the ship. These readings are then turned into an accurate map.

The Making of an Ocean

Could the Red Sea become an ocean? Scientists think it's possible. The Red Sea sits at the point where three tectonic plates have been pulling Africa and the Arabian Peninsula apart for the past 10 to 20 million years. If this motion continues at the rate of 1 cm (0.4 in.) a year, in about 400 million years the Red Sea will be as wide as the Atlantic is today.

Red Sea

Seamount

Guyot

Abyssal Plain

Continental Shelf

Continental Slope

Mid-Ocean Ridge

Trench

What Covers the Floor?

Much of the rocky ocean floor is blanketed by a layer of soft **sediment** hundreds of meters thick. This thick covering has built up over millions of years. It is composed mostly of tiny bits of **inorganic** material, like mud and sand, and the skeletons of microscopic marine plants and animals *(magnified above)*.

Would You Believe?

A Unique Map

In 1977, geologists and map-makers Marie Tharp and Bruce Heezen *(right)* unveiled this amazing map of the ocean floor *(below)*. It was an incredibly clear view of a landscape that, until then, had been more remote and unknown than the surface of the moon. It took 30 years to create the map. Heezen spent three years at sea collecting information about the sea-floor. Back in their laboratory, Tharp used Heezen's findings and other information gathered over the past 100 years to produce her beautiful, highly detailed drawings. In the course of their work, Heezen discovered the global mid-ocean ridge and Tharp correctly concluded that it was caused by fault lines along the edges of tectonic plates.

Ocean Water

I f all the water in the world's oceans evaporated, it would leave behind a layer of mineral deposits about 13 m (43 ft.) thick. The most common of these minerals is sodium chloride, or salt, which gives seawater its salty taste. If you've ever swum in the ocean, you probably noticed the salty taste of the water, or the salt left on your skin after the water evaporated.

Billions of years ago, rivers flowing into the newly formed oceans carried small quantities of salt and other dissolved minerals. An ocean's **salinity** refers to the total amount of these minerals dissolved in the water.

The salinity of the major oceans has stayed the same for thousands of years. Rivers still carry minerals into the ocean, but once there, they don't all stay in the water. A portion of them sink to the bottom to mix with the **sediment,** some are released into the air, and others are taken in by plants and animals.

Floating in the Dead Sea

The Dead Sea in Israel is the perfect place for people who can't swim to go swimming. We float in water because our body is less **dense** than the surrounding water. It is easier to float in salt water than in fresh water because salt makes the water denser. The salinity of the Dead Sea is so high—about 10 times that of the ocean—that it's impossible to sink! As soon as you wade waist deep into the water, your legs float out from under you. And be careful not to get any drops in your eyes or on your tongue. It stings terribly and tastes awful.

Why Is Ocean Water Blue?

S unlight is composed of many different colors. You can see these colors when you look at a rainbow. When sunlight strikes the surface of the ocean, all the colors are absorbed by the water except for blue. The blue rays are absorbed the least and are reflected back toward our eyes. That's why we see blue.

The ocean can appear to be many different shades of blue (above). Clear, shallow water close to a shoreline with a sandy bottom looks light blue or turquoise. The open ocean is dark blue, or sometimes almost gray green if it contains lots of microscopic **marine** life. Where sea grasses and **algae** grow, the water appears almost black.

Where's the Salt?

When salt water freezes, most of the salt doesn't freeze with it. Where does it go? The salt stays behind in the unfrozen water, making it even saltier. Salt water freezes at a lower temperature than fresh water. So the saltier the water, the less likely it is to freeze when it's really cold. That's good for animals who live in and around the Arctic Ocean *(above)*. No matter how cold it gets, there's always unfrozen water under the floating ice for fish and other marine life to live and hunt in.

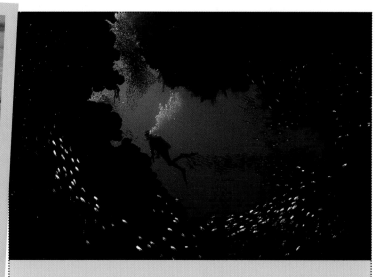

What Is the Bends?

The bends is a serious condition that affects divers. It causes intense muscle pain, dizziness, nausea, and sometimes even death. In deep water, gases in the diver's body are put under a lot of **pressure**. If the diver surfaces too fast, the pressure is suddenly released. Gas bubbles form in the blood, much like in a can of soda that is opened immediately after it has been shaken. To avoid the bends, divers must come up slowly with frequent pauses.

Salt Harvest

Much of the salt we use today comes from the ocean. Shallow artificial lakes called salt pans, like these in Thailand, are flooded with seawater. The sun evaporates the water, leaving the salt behind. Salt has been a highly valued item throughout history. Roman soldiers were even paid in salt! The word "salary" comes from the Latin word meaning "salt money."

Pollution: A Serious Threat

Oil poses a hazard to all marine life. Seal and salmon populations were severely affected by the wreck of a tanker off the coast of Scotland in 1993 *(below, right)*. Sometimes oil-covered birds, like the king penguin below, can be captured and cleaned. But oil spills are not the only threat to ocean water. If not properly handled, household chemicals, lawn fertilizers, and pesticides can end up in streams and rivers and, eventually, the ocean.

Speeding through Water

Bermuda

Perth

Sound travels faster through water than through air. At certain depths, a condition called a sound channel can carry it much farther. In 1960, the sound of an explosion off the coast of Perth, Australia, was picked up by scientists in Bermuda. In a little under four hours, the sound had traveled 19,000 km (12,000 mi.)!

Ins and Outs of Waves

Ocean water is constantly in motion, from small ripples to huge, storm-tossed seas. Most waves are created by wind moving across the ocean, stirring up the surface. Normal ocean waves run about 1.5 to 3 m (5 to 10 ft.) high. But during a violent storm with high winds, waves can reach more than 20 m (65 ft.) in height.

The size of a wave depends upon three things. The first is how hard the wind is blowing. Second is the length of time the wind is in contact with the water. And third is the distance the wind travels over the water surface. This is called the fetch. The longer the fetch, the bigger the wave. Waves can travel clear across the open Pacific before striking land.

In 1933, a wave estimated at about 34 m (112 ft.) high nearly sank the navy tanker USS *Ramapo,* which was traveling between the Philippines and California during a fierce hurricane. This wave holds the record as the highest sea wave formed by the wind.

Measuring Waves

As a wave moves up and down, you can see two distinct parts: the crest, or top, and the trough, the lower part. Oceanographers measure waves in several ways. A wave's height is the distance from the tip of the crest to the bottom of the trough. Wavelength is the distance between one wave crest and the next. A wave period is the amount of time it takes two crests to pass a fixed point, such as a pier or a ship.

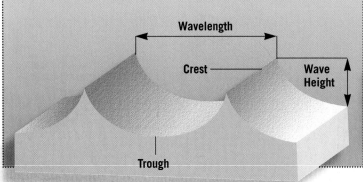

Wavelength

Crest

Wave Height

Trough

The Ocean in Motion

When fans do "the wave" at a sports event *(right),* they move up and down without changing seats. The water in an ocean wave does about the same thing. The wind pushing the crests and troughs across the surface makes the water appear to move sideways, but it is really just going up and down.

Hidden in a deeper layer under the ocean's surface are immense, slow-moving waves we can't see. They are called internal waves and can be as high as 90 m (300 ft.). Oceanographers think they are caused by tidal currents or underwater avalanches.

Shoreline

Wind Direction

Surface Waves

Internal Waves

A Gallery of Waves

Surf's Up!

A windsurfer makes a quick getaway from this monster wave on the Hawaiian island of Maui. These waves can be as tall as a five-story building and come barreling ashore at about 40 km/h (25 mph). Their crashing sound can be heard miles away.

The shape and size of a wave is determined by the wind. Sudden, short gusts stir up chop waves, which smooth out when the wind dies down. A steady wind blowing across a large area produces long, rolling waves called swell. Strong storms blow water right off the waves' crests, creating foamy whitecaps. The shallow ocean bottom near shore slows a wave's trough while the crest moves forward and topples over. This is called a breaker *(top)*.

Chop

Swell

Whitecap

What's Sea Foam?

Have you ever noticed frothy bubbles washing up on a beach or floating on the surface where a boat has churned up the water? These tough, long-lasting bubbles are called sea foam. Sea foam comes from **organic** matter in the ocean. Any substance produced by a living organism is considered to be organic matter.

Tsunamis: Watery Giants

Tsunamis are often called tidal waves, but they have nothing to do with tides. Underwater volcanic eruptions, earthquakes *(diagram below)*, or landslides create these enormous waves. Their name in Japanese means "harbor wave" and they occur most frequently in the Pacific. The highest wave ever ridden was a tsunami. In 1868, a Hawaiian man named Holua saved his life by successfully riding a tsunami that was "perhaps 15 m (50 ft.)," according to *The Guinness Book of World Records*.

The Power of Waves

When a wave breaks against the shore *(above),* it releases a lot of energy. In a single day, all the waves crashing onto all the world's beaches produce the force of a 50-megaton hydrogen bomb. Combined with wind and tides, huge waves slamming against the coastline can greatly change its shape. On New York's Long Island *(below),* a stranded beach house and exposed fire hydrant clearly show the damaging consequences of wave action.

What's Sand?

Sand is that wonderful, fine-grained stuff that's great for building castles or just wiggling your toes in. It is made of a variety of materials that are ground down by wave action into tiny particles. The pink beaches in Bermuda are the skeletal remains of microscopic **marine** animals. The yellow sand of California is made mostly of powdery granite. Many tropical beaches have bright white sand made of crushed seashells, coral, or echinoderms such as sea urchins *(pages 50-51).*

Fabulous **Features!**

Erosion and Deposition

Waves are continuously changing the world's shorelines through two forces. Erosion is the wearing away of rock or soft **sediment.** Deposition refers to how wave action moves this eroded material to another place. The shifting sands of a sand bar in Australia *(above)* show deposition hard at work.

A sandy coastline is altered much more quickly than rocky cliffs. But over a long period of time even the hardest rock can be carved into fantastic shapes by the force and motion of waves like this one on the coast of California *(below).*

A Shore's Sandy Defense

Barrier islands are the last line of defense between the mainland and the pounding ocean surf. These long, narrow bodies of sand run parallel to the seashore. Shaped and reshaped by wind, waves, currents, and storms, they are found only off gently sloping coasts. The longest chain of barrier islands runs more than 2,400 km (1,500 mi.) along the East Coast of the United States.

Clever Snail Grabs a Ride

Bobbing along on the ocean waves off the coast of Bermuda, the violet snail has evolved a unique way to stay afloat. It creates a raft of bubbles by trapping air in the cupped end of its foot, then covering it with **mucus.** The snail drifts along the surface hoping to bump into a tasty jellyfish *(above),* its favorite food. But the snail has to be careful. If it loses its raft, it will sink and die.

Would **You** *Believe?*

Black Sand

No, this beach in Hawaii has not been polluted by an oil spill. The sand is actually black! These volcanic islands are made of black lava. Over time, large stretches of the lava have been ground up by wave action to create unique black beaches.

Highs & Lows of Tides

Both the sun and moon pull on the earth with a force called gravity. Although the sun is many times larger than the moon, the moon is much closer, so its pull on the oceans is twice as strong as the sun's. In addition, the effect of the moon's path around the earth causes a second force that combines with gravity to produce the twice-daily rise and fall of ocean waters that we call **tides.**

The distance between the water level at high and at low tide is called the tidal range. It varies around the world depending on the shape of the ocean basins and coastlines. The normal range is about 1 m (3 ft.), but it can be as small as a few centimeters. The highest range in the world—15 m (50 ft.) between high and low tide twice a day—occurs in the Bay of Fundy on the Atlantic coast of Canada.

What Causes Tides?

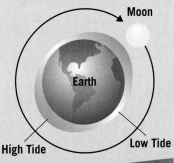

The moon's strongest pull is on that part of the ocean directly beneath it. Its gravity creates a bulge in the surface of the ocean. This bulge is high tide. On the opposite side of the earth, there is a similar bulge, which forms another high tide. It is caused by the moon's orbit around the earth, which pushes the ocean water away from the surface. The two low tides are the areas in between the bulging high-tide areas. Approximately once every 24 hours each spot in the oceans normally experiences two high and two low tides.

Low tide in Erguy, France.

Spring and Neap Tides

Spring Tide

Moon · Sun · Earth

Neap Tide

Twice a month, during the new and full moons, the sun, moon, and earth are in a straight line. The combined pull of gravity from both the sun and moon causes higher than normal tides called spring tides. But they have nothing to do with the season. The name comes from an Old English word meaning "to jump up."

The sun and moon are also at right angles to the earth twice each month. When that happens, the gravities of the sun and moon work against each other, and the tides rise and fall less than at any other time of the month. This is called a neap tide, from the Old English word meaning "napping."

Beware the Whirlpool!

Whirlpools can be found all over the world, wherever strong tides collide with surface currents or flow over a shallow, uneven seafloor. This whirlpool *(above)* is located in the Mediterranean Sea between the island of Sicily and the Italian mainland. It measures almost 2 m (6 ft.) across at its center. According to local fishermen, the whirlpool is large enough to swallow small boats.

Life in the Tidal Zone

Many **marine** animals have evolved strategies to survive the daily ins and outs of the tides *(pages 92-93)*. The white mussel buries itself in sand or mud, then pokes two tubes up into the water at high tide to get food and oxygen. Scientists think the mud fiddler crab has an internal "tidal clock" that times the rise and fall of the tides. When the tide retreats, the crab comes out to eat. It then returns to its burrow before the tide rushes back in.

White Mussel

Mud Fiddler Crab

Capturing Tidal Energy

Pollution-free energy can be produced from the power of strong tides. The first tidal-power station was built in 1966 across the mouth of the Rance River in northern France. The flowing water of the rising and falling tides drives 24 turbines located below the water level. The turbines are attached to generators that convert this tidal energy into electricity.

A tidal-power station in France.

Currents Rivers in the Ocean

Great rivers of water, called currents, move huge amounts of water along the surface and deep within the ocean. Often moving at walking pace, surface currents can be hundreds of kilometers wide and hundreds of meters deep. They are created by strong, steady global winds that blow in one direction. The earth's rotation also affects their speed and direction.

Some currents found deep in the ocean are caused by a difference in **salinity** and temperature. The cold, **dense,** salty water found in the polar oceans sinks as it moves slowly toward the equator. It then rises as it warms, replacing the heated, tropical ocean waters that are moving back toward the Poles. This continuous circulation can be a long process. The frigid bottom water moves so slowly that it may take up to 2,000 years to return to the surface.

North Pacific Gyre

North Atlantic Gyre

South Pacific Gyre

South Atlantic Gyre

Indian Ocean Gyre

North Atlantic Gyre

Earth's Rotation

South Atlantic Gyre

Currents of the World

There are thousands of currents of varying sizes that flow along the ocean's surface. They travel at speeds ranging from 16 to 160 km (10 to 100 mi.) a day. As they move, the currents are deflected off the

The earth's rotation affects how currents move. North of the equator the earth's spin forces them to the right. To the south they are forced to the left.

continents, causing them to flow and merge into five giant circles of moving water called gyres (pronounced "jires").

At the age of 14, Subaru Takahashi took on the voyage of a lifetime. On July 22, 1996, he set out alone in his 9-m (30-ft.) sailboat *Advantage* from Tokyo Bay, Japan. He was bound for San Francisco— 9,600 km (6,000 mi.) away! After a power failure on his boat, he could rely only on the wind and the ocean currents to carry him to his destination. On September 13—almost two months after he set out— Subaru arrived safely in San Francisco Bay, the youngest person ever to sail solo across the Pacific Ocean.

Parallel Currents

The lighter blue, fresh water flowing from New Zealand's Clarence River remains separate from the darker waters of the Pacific Ocean. Since fresh water is less dense than the salty ocean water, they don't instantly combine. Depending upon how strong the currents are along a coast, river water may take months or even years to mix with the water of the open ocean. Variations in water temperature also keep some currents from blending.

Shoreline Currents

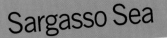

Sargasso Sea

North America

Europe

Sargasso Sea

Africa

South America

In the calm waters at the center of the North Atlantic Gyre is a region of the ocean called the Sargasso Sea. It is the only sea bordered not by land but by ocean water *(map above)*. Named after its abundant floating seaweed, sargassum *(top left)*, it is nearly the size of the continental United States. Many small **marine** creatures live in the drifting seaweed. Animals like the sargassum crab *(left)* have evolved camouflage to blend perfectly with the seaweed's yellow brown color.

Open Mouth, In Flows Food

The constant movement of currents provides transportation for some animals and brings a plentiful supply of food to others. Jellyfish *(pages 38-39)*, for example, drift along at the whim of the ocean's currents in search of a meal. California sand dollars, a type of flattened sea urchin *(page 51)*, sit in the sand without moving, the edge of their shells angled to face the oncoming currents. The moving water delivers food right into their mouth.

Have you ever dropped your towel on the beach and raced into the water for a swim, only to notice a short time later that you have moved down the beach from your towel? You were caught in a shoreline current, a type of current that runs parallel to the beach.

On some beaches, after a wave breaks, the water flows back out to sea underneath the oncoming waves, creating a rip current *(red arrows above)*. The strength of this current depends upon the size of the waves and the amount of water they are throwing up on the shoreline. Rip currents, sometimes called undertow, can be very dangerous because they can pull a swimmer out to sea.

Weather and Climate

The ocean's surface is like a giant sponge that soaks up the sun's heat in the summer and releases it in winter to warm the earth. The **dense,** salty ocean water can store much more heat energy than the atmosphere above it. As the heat is released back into the air above the ocean, it adds moisture and creates wind. And the wind in turn makes waves and currents that move the stored heat around the world.

On a daily basis, heat and moisture from the ocean are responsible for our weather, including our fiercest storms, known as hurricanes or typhoons. Over time, the ocean even controls the earth's climate. Without vast oceans to regulate the temperature, the earth's surface would be much hotter during the day and much colder at night.

What's Sea Fog?

The rusty-red twin towers of the Golden Gate Bridge are the only visible sign of the famous landmark on San Francisco Bay. Each summer, huge banks of fog roll in from the Pacific Ocean onto this part of the California coast. Sea fog forms when low-lying warm, moist air cools as it moves over cold sea-water. The moisture **condenses** out of the air, creating a dense, earthbound cloud.

How Nature Recycles Water

Water moves in a continuous cycle between the ocean, the atmosphere, and the land. Most of the moisture in the atmosphere comes from the ocean. The sun's warmth evaporates water from the ocean surface, adding water **vapor** to the air. As the air condenses, heat is added to the atmosphere, clouds form, and rain falls. Most of this moisture returns directly to the ocean. Some falls as rain or snow over land, where rivers take the water back to the ocean to start the cycle all over again. This cycle is known as the water cycle.

When Weather Goes Wild

Captured in a 1996 satellite photo *(top right)*, Hurricane Fran beats a destructive path toward the eastern coast of Florida. Near the equator, warm surface waters of the open ocean put lots of moisture into the air. This hot air rises rapidly, condensing to form huge clouds that spin around a central "eye," or calm low-**pressure** area.

The name hurricane comes from the Caribbean Indian word *hurrican,* meaning "evil spirit." These violent, spinning storms have different names in different parts of the world. In the western Pacific they are called typhoons. In countries bordering the Indian Ocean, they are called cyclones.

Hurricanes are known for their flooding rains, powerful surf *(bottom, right),* and high winds of at least 119 km/h (74 mph). Once the storm hits land or passes over cooler water, it gradually loses its strength.

Crazy Currents

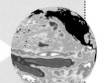

Nov. '97

July '98

Every two to 10 years, surface currents in the Pacific Ocean around the equator change direction. Instead of their normal westward flow they start moving east, pushing warm surface water *(white on top globe)* toward South America. Since it happens around Christmastime, it is known by the Spanish term *El Niño,* meaning "The Christ Child." Eight months later *(bottom globe),* there is little left of the pool of warm water, which signals the end of El Niño. This crazy shift in currents causes disastrous events worldwide: heavy rains and floods in some areas, drought and dust storms in others. In Australia's Great Barrier Reef, the water grew so warm during one El Niño that the coral turned a sickly white *(below).*

A Current Changes Climate

Did you ever imagine that palm trees could grow in Scotland? This garden in Galloway *(left)* on the south-west coast of Scotland enjoys an unusually mild climate for that part of the world. It's all due to a warm ocean current called the Gulf Stream that starts thousands of miles away in the Caribbean and Gulf of Mexico. Carrying 100 times more water than all the world's rivers combined, it sweeps warm tropical water northward at about 100 km (60 mi.) a day. Without the Gulf Stream, Scotland's climate would be as cold as northern Canada's.

North America

Gulf Stream

Europe

Gulf of Mexico

Africa

South America

Algae and Seaweed

Most of the ocean's plants are **algae,** a name familiar to anyone who has swum in a pond or cleaned a fishtank. Not all algae are tiny green specks, though. Most algae **species** are microscopic, but the family includes a gigantic form of seaweed called giant kelp that grows to 50 m (165 ft.). Seaweed is the name given to many forms of algae that live in the shallow edges of the ocean.

Like most other plants, algae use **photosynthesis** to convert sunlight to food. Because sunlight cannot penetrate deep water, algae grow only near the surface and in the shallows. Different colored seaweed lives in different depths of water *(right),* creating a multicolored underwater seascape. One form of red seaweed can form hard, crusty structures resembling coral reefs. It is appropriately called coralline seaweed, or algae.

Types of Seaweed

Let's Compare

Green Algae
About 70 percent of seaweeds live between the low- and high-**tide** lines. Sea palms *(right)* and other green algae grow near the top of the high-tide mark on rocky shores. They need fairly bright sunlight to live.

Brown Algae
Brown seaweeds like kelp *(right)* get their color from a **pigment** called fucoxanthin. Fucoxanthin can absorb light in slightly deeper water, and so brown algae grow lower in the intertidal zone.

Red Algae
Red algae, such as *Rhodymenia (right),* can live in relatively dim light. Red seaweeds grow deepest in the intertidal zone. Thus, intertidal zones are striped: green, brown, red!

People   Seaweed Lady

Like many a scientist before her, Judith Connor has come to love the life forms she studies. In her case, those life forms are not fuzzy mammals, but algae. "Algae are incredibly beautiful," says Connor, "like wild-flowers. They're so many different shapes, colors, and sizes."

Connor's colleagues call her the Seaweed Lady. A more scientific title is phycologist—one who studies algae. Connor dons scuba gear and studies the giant kelp beds off the California coast. Connor gets a kick out of her work, as you can perhaps tell by the kelp crown she's wearing in this picture.

Shown at her side is her son and helper, Alexander Gregory. His nickname? Al G, what else?

Seaweed Parts

Algae are classified as simple plants. They lack the roots, stems, leaves, and flowers of so-called higher plants, but they do have special body parts to help them float and stay put in the water. Leaflike blades absorb **nutrients** and sunlight. Gas-filled floats keep the blades bobbing upward *(right, top).* Rubbery "stems" are called stipes. Sea-weeds don't have roots, but they do have holdfasts *(right, bottom),* which allow them to grip under-water rocks.

Oxygen Producers

Gas bubbles up from an underwater mat of green algae *(left)*. The gas is oxygen, given off as a waste product as the plant uses sunlight for energy, and nutrients from the surrounding water, to make food. The oxygen doesn't go to waste! It is, of course, the element animals need for life—the one you absorb with every breath. **Marine** algae contribute much of the earth's oxygen gas.

Bet You Can't Find Me

Unquestionably, algae are useful. They produce vital oxygen gas, and they feed billions of sea creatures and are home to many others. They are an important source of human nutrition in many cultures, too. But, can you wear algae? Some crabs think so. The spider crab *(above)* deliberately snips bits of seaweed with its claws and sticks them onto its limbs and back. Fine, Velcro-like bristles cover much of the crab's surface, providing the "glue." The crab doesn't do it to look good, but rather to camouflage itself from **predators**.

True-Blue Friends

What do giant clams and gardeners have in common? They both grow plants for food! Living within the clam's **tissues** are millions of microscopic algae, which give the clam its beautiful blue color. The clam provides a sunny home for the algae. In return for this hospitality, the algae use photosynthesis to turn the sun's energy into food for both them and the clam. This kind of happy relationship between two creatures is called mutualism, a type of **symbiosis.**

What's Sea Grass?

Sea grass is the ocean's only flowering plant. It grows in shallow, calm waters. Eelgrass *(below)* and turtle grass are two varieties. Like their land relatives, these ocean plants produce pollen, but it is transported by water instead of by air. Sea grasses lack the color and flash of a coral reef, but they play an equally important role. They keep bottom **sediment** from washing away and provide food for some animals and shelter for many others.

Kelp Forests

Sequoias of the seaweed world, giant kelp form dense forests *(right)* off the shores of several continents. They need a hard surface on which to attach themselves; clear, cool, **nutrient**-rich water; and a place where currents are not too strong. In return, they provide a **habitat** for numerous other **species** of plants and animals. About 1,000 species of kelp are found worldwide. They range in size from a modest 31 cm (12 in.) to a whopping 50 m (165 ft.) in length, but all kelp species are similar in looks and life history. A rootlike holdfast anchors them to underwater rocks. Air-filled floats keep the blades at the surface, closest to the sun. Between holdfast and blade, rubbery stipes whip about in the waves, bending where a more rigid trunk would snap.

Where in the World?

Kelp forests are found around the world in shallow coastal waters.

Harbor Seal

A shy, speckled harbor seal swims through a kelp forest searching for a tasty meal of fish. Seals often visit kelp forests in search of food and shelter but aren't the year-round residents otters are.

Seaweed: A Useful Crop

In Asia, Africa *(right)*, and especially in Japan, people harvest kelp and other seaweed for use in soup, sushi, and many other dishes. Westerners eat a lot of seaweed—mostly without knowing it! A kelp ingredient called algin is used in products from ice cream and toothpaste to house paint. Algin's usefulness comes from its ability to keep tiny particles in **suspension** in a liquid. Without it, a milk shake would be more like curdled milk instead of smooth and creamy.

Sea Urchins

Like termites in a forest, a prickly army of purple sea urchins moves through the kelp forest munching every speck of kelp in its path. Sea urchins have always been part of the kelp forest **ecosystem.** But their populations have exploded in areas where their natural **predators**—principally, the sea otter—have disappeared. The presence of too many urchins turns a lush kelp forest into a dreary urchin barren *(above).*

Sea Otter

King of the kelp jungle, a whiskery sea otter snoozes among the kelp blades. A sea otter may spend its entire life in the kelp forest. Mothers park their babies at the surface, twined in a kelp safety belt to keep them from floating away. Adults dive to the bottom for abalone and sea urchins to eat. Sea urchins have a tremendous appetite for kelp, so by eating them, otters help keep the ecosystem in balance.

Snails

Jeweled-top snails move slowly along a kelp stipe, stripping a surface layer of dead plant material. The snails feed without damaging the plants. Their filelike teeth cannot pierce the tough kelp.

Garibaldi

The male garibaldi is a stay-at-home dad. He builds a nest, cleans it, and aggressively defends it from any intruders. When eggs are laid, he carefully watches over them until they hatch two weeks later.

Plankton Ocean Drifters

I f there were a prize for contributions from the plant **kingdom,** the winner would be clear: a humble mass of mostly one-celled plants that float near the ocean's surface. Scientists call these tiny plants the meadows of the sea, and they form the base of the **marine** food chain on which all other sea life depends. Tiny animals float along with the plants. Together, this mass of marine life is called **plankton,** a Greek word meaning "drifting." Phytoplankton, or plantlike plankton, must live near the surface of the oceans because they need the sun's energy to make food. Their tiny animal companions—called zooplankton—feed on the phytoplankton. Directly or indirectly, almost every living thing in the ocean owes its life to plankton.

How Big?

Plankton

Plankton come in all shapes and sizes. The smallest are **bacteria** so tiny that scientists find as many as two million in a single teaspoon of seawater! Although bigger, all phytoplankton are small—smaller than a human hair is wide. Diatoms *(below),* the most plentiful phytoplankton, have beautiful shapes and shell-like casings. The shapes help keep them from sinking. Most zooplankton are very small, but some jelly-fish *(right)*—considered to be plankton because they drift rather than swim—grow to 9 m (30 ft.).

Jellyfish Giant

Let's Compare

Phytoplankton and Zooplankton

Phytoplankton and zooplankton are as different as, well, plants and animals. Phytoplankton stay within a few hundred meters of the surface. They are one-celled, but some hook together in chains. They make food with energy from the sun. Zooplankton drift up and down as much as 900 m (3,000 ft.) every day. They look like tiny animals, and they eat other plankton. Some are multicellular. Many kinds of zooplankton spend only part of their life as plankton *(next page).*

Phytoplankton's strange shapes help keep them afloat.

Zooplankton look like the tiny marine animals they are.

One-Celled Diatoms

Part-Time Plankton

Not all plankton spend their entire life drifting in the water. Many zooplankton are part-timers: drifters in their early, larval forms, and swimmers or crawlers as adults. This ghostly, spindly-legged little creature *(right),* for example, is the larval stage of a slipper lobster. The more familiar looking adult is shown in the smaller photo. Other animals with a planktonic stage as youngsters include some fish, crabs, barnacles, worms, clams, snails, and sponges.

Phytoplankton's Global Production

No, the oceans haven't suddenly gone Technicolor, just computer enhanced. A satellite photograph *(above)* shows the distribution through-out the world's oceans of the microscopic marine plants called phytoplankton. Special instruments on board the satellite can "sense" the presence of chlorophyll, a green substance found inside all plants. Plants need chlorophyll for **photosynthesis.** In the map above, the bright yellow, orange, and red areas near the coasts indicate lots of phytoplankton.

Would **You** *Believe?*

Diatom Clues Solve Crime

Four valuable hubcaps were stolen from an antique car in England. The police had a suspect but no way of proving that he stole the hubcaps, which were now on his car. The police examined the polish on the hubcaps. Many polishes contain diatoms because their glasslike silicon coats make them mildly abrasive. Sure enough, the diatoms in the hubcap polish matched those in the polish on the antique car. Bingo! Crime solved.

Plankton Close up

Catching Food

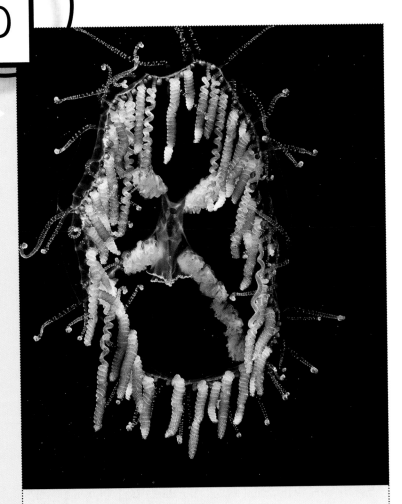

Trolling for smaller zooplankton, the cigar jelly *(below)* trails long tentacles through the night sea. It traps captives and hoists them up to its mouth for a meal. By day, the jelly coils the tentacles tightly *(right)*. Other zooplankton are filter feeders—they simply drift through **plankton**-rich waters with their mouth open. Membranes in their body strain out food. Krill, tiny shrimp-like plankton, have minute hairs on their limbs that sweep food into their mouth. Krill are the main food of baleen whales *(pages 88-89)* and animals that live in the ocean around Antarctica *(pages 104-105)*.

Staying Afloat

Phytoplankton must stay in the sunny top layer of the ocean, where they can make food, or they will die. Leaflike, spiral, and other shapes help them stay afloat. Zooplankton don't need sun, but they must live where the phytoplankton do. Many have tails that help sweep them in the direction they need to go. The by-the-wind sailor, a jellyfish *(right)*, relies on a sail-like fin and a skirt. The sail catches the wind, and the skirt provides stability.

What's a Red Tide?

Green scum on water is so common it usually doesn't merit a second glance. Red scum, like that in this photo of waters off Baja Peninsula, Mexico *(right)*, is another story. A sudden population bloom in a species of dinoflagellate, a phytoplankton, causes the red stain, called a red tide. Red tides can be very dangerous. Tiny amounts of poisons in each plankton build up in plankton feeders. Fish and shellfish can be poisoned—and so can the humans who eat them. The poison is 50 times stronger than rattlesnake venom.

How Big?

Largest Bacteria

Scientists announced with great fanfare, in April 1999, that they had discovered a new kind of **bacteria,** one of astonishing size. Found living under 90 m (300 ft.) of water off southern Africa, the incredible creature is—are you ready?—a little bigger than the period at the end of this sentence. Well, perhaps this is a creature only a scientist could get really excited about. The new bacteria are the giants of the microscopic world, 500 times the size of an average bacterium.

The Best Defense: Spitting at the Enemy

This underwater photo catches a decapod shrimp doing its utmost to escape a bristlemouth fish. As the fish approaches open-mouthed, the little shrimp spits out a blinding cloud of luminescent liquid into the fish's face. At the same time, the shrimp flips its tail forward, hard, and scoots backward to safety. The shrimp is a kind of zooplankton—though it can swim, it's not a strong swimmer. That's why glow-in-the-dark vomit is its best defense!

Glowing for a Row

Swimming with dolphins is a big hit at many resorts. Who would imagine that swimming with plankton could be even more thrilling! Anyone who has ever seen a firefly blink its light has witnessed a phenomenon called **bioluminescence.** This is light made by a living thing, created by chemical reactions inside special cells. Some kinds of plankton have bioluminescence, too, though no one is sure why. The tiny animals *(inset)* sparkle when the waters they float in are disturbed. When there are millions of bioluminescent plankton, as there are in a bay off Vieques Island in Puerto Rico, a nighttime row can be magical.

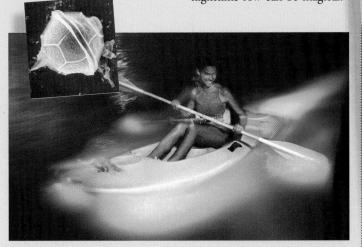

Strange But TRUE!

Green Iceberg

Your average, *Titanic*-sinking iceberg is bluish white, formed when chunks of ice break free from a glacier. The undersides of some of those chunks are made up of frozen seawater, as well as compressed snow. If that seawater was rich in plankton, the iceberg's underside may be green. When that iceberg capsizes, as icebergs will, the unlikely result is—a green iceberg *(above)!*

What Is a Food Web?

In a kelp forest *(pages 26-27)*, sea plants called kelp use the sun's energy to make food. Sea urchins graze on kelp. Sea otters eat sea urchins. Killer whales hunt sea otters. Biologists call this a food chain, millions of which exist on earth. However, the idea of a chain implies a straightforward series of links. Biologists prefer the name food web, which better describes the complicated tapestry of eater and eaten. Sea urchins don't eat just kelp. They devour other **algae** species, too. Sea otters eat abalone and other shellfish. Killer whales vary their diet with seals and sea lions, penguins, and other **prey.** Everything is connected; all the **species** are interdependent.

What's Marine Snow?

As a spider spins a web to strain food from the air, so the tiny zooplankton called a larvacean (related to salps and sea squirts) builds a gelatinous "house" around its body to strain food from the water. In its lifetime—just a few days—the larvacean may build and discard 40 houses. As the houses sink, **plankton** and other material gets stuck to them like bugs in a spider web. Millions of drifting houses may descend through the water, creating the snowstorm of specks called **marine** snow. In some parts of the ocean there may be as many as 26,000 houses per cubic meter of water. This blizzard of marine snow brings food to the ocean depths.

Ocean Food Web

In the ocean food web, the food produced by the phytoplankton is passed to the zooplankton and then to **predators** of increasing size. The drawing below shows the elements of the web. The photo above captures the elements in action. The brown-and-white band is where waves are breaking on a beach.

The wavy black band is made up of millions of tiny fish called anchovies that school along the Australian coast during winter. The anchovies find plankton there. In the deeper blue green waters, sharks, whales, and predatory fish attack the anchovies in a feeding frenzy.

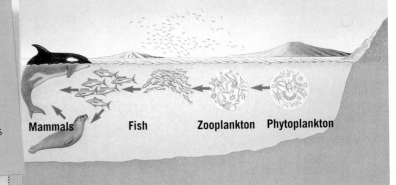

Mammals   Fish   Zooplankton   Phytoplankton

Habitable Zones of the Ocean

Like a rain forest, the ocean has distinct vertical **habitats,** and most creatures live within one zone. Most ocean life is found in the top 180 m (600 ft.), where water is warm, sunny, and constantly moving. Plants cannot grow in deeper waters, so zooplankton and other plankton feeders must live here, too.

From 180 m (600 ft.) to 1,000 m (3,300 ft.) lies the dimly lit twilight zone. Water temperatures here dip as low as 5°C (41°F). Giant squid—at up to 20 m (66 ft.) they are earth's largest **invertebrates**—live here. Sperm whales descend to the twilight zone to prey on squid but return to the surface for air. Many fish here have glittering **bioluminescent** spots along their body and tail. The little lights help them attract prey and mates.

From 1,000 m (3,300 ft.) to 4,060 m (13,200 ft.) water **pressure** mounts, rising as high as 1,000 times surface pressure. There is very little food here.

From 4,060 m (13,200 ft.) to 6,030 m (19,800 ft.) are the abyssal plains, which cover almost half the deep-sea floor. Thick mud covers this cold, dark surface.

Sunlit Zone

Twilight Zone

Dark Zone

Abyss

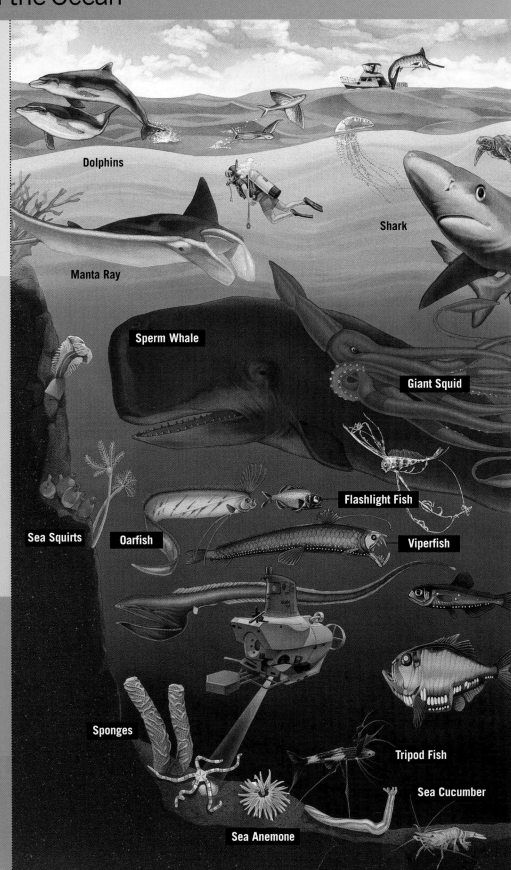

Dolphins

Shark

Manta Ray

Sperm Whale

Giant Squid

Flashlight Fish

Sea Squirts

Oarfish

Viperfish

Sponges

Tripod Fish

Sea Cucumber

Sea Anemone

What's an Invertebrate?

Ninety-five percent of all **animals** alive today are missing something you have: a backbone. Animals without a backbone are called **invertebrates.** You are part of a much smaller group known as **vertebrates** because you have a spine made up of bones called vertebrae running down your back. Your spine helps support your body. Invertebrates don't have this support system, but they've developed other support strategies. A clam has a hard outer shell, a crab has a sturdy outer covering, and a jellyfish is firm when it's filled with water.

There are more invertebrates living on earth—on land and in the ocean—than any other animal. Insects are the most common. Scientists believe there may be as many as 4.5 million different species of invertebrates, but only about half have been discovered.

Fast FACTS

Scientists organize the animal **kingdom** into groups of animals based on the characteristics they share. Each major group is called a **phylum.** More than 25 of the animal phyla include **marine** invertebrates. Here are some of the major ones:

Porifera 5,000 species of sponges.

Cnidaria 9,000 species, including jellyfish, sea anemones, and corals.

Platyhelminthes Commonly known as flatworms; about 3,000 species live in the ocean.

Annelida More than 12,000 kinds of worms with segmented bodies; known as the "earthworms of the sea."

Mollusca 50,000 species, including clams, slugs, octopuses, and squids.

Arthropoda The biggest phylum in the animal kingdom; more than 750,000 species—and maybe as many as two to three million; 38,000 are **crustaceans** such as lobsters, shrimp, and crabs.

Bryozoa 4,000 species of tiny animals that usually live in **colonies** attached to rocks or other hard surfaces.

Echinodermata About 6,000 species of spiny-skinned animals like sea stars, sea urchins, and sea cucumbers.

Chordata Fewer than 2,000 species in this phylum are invertebrates, such as sea squirts and lancelets. Vertebrates make up the rest of the phylum. Humans are **chordates.**

This picture looks like a lovely garden of colorful marine plants. In fact, many of these "plants" are invertebrate animals. If you look very closely, you will see a number of feather stars—relatives of sea stars—a fire urchin, sponges, hard coral, and a sea cucumber. As this gallery of invertebrates shows, the appearance of animals without a backbone couldn't be more diverse.

A World of Difference

Squid

The squid *(pages 44-45)* is a smart hunter and a fast swimmer. Without an external shell (like its relatives the clam and snail), this **mollusk** relies on its sharp vision and intelligence to survive.

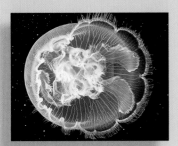

Jellyfish

The slow-moving, gelatinous mass called a jellyfish *(pages 38-39)* may seem like an easy target for **predators**. But its stinging tentacles, common to all cnidarians, are great weapons for hunting and defense.

Crab

The crab's jointed legs and protective outer **skeleton** make life without a backbone a little less frightening. Also helpful are the claws and pincers these arthropods use for defense and attack *(pages 48-49)*.

Sea Urchin

Looking like a big underwater porcupine, the sea urchin *(pages 50-51)* protects its soft body with a covering of sharp spines. The spines of these echinoderms sometimes contain poison.

Sea Squirts

They look simple, but sea squirts are the invertebrates that are most closely related to humans and other vertebrates. That's because they share some of the same features early in their development *(page 53)*.

Symmetry refers to the way an animal's body is organized around a central point or line. Some animals, like sea stars *(top)*, form a circle around a center point. This is radial symmetry. A crab *(center)*, like a person, has bilateral symmetry. If a line were drawn down its middle, each half would be a mirror image of the other. Some animals, such as most sponges *(bottom)*, aren't organized around a central point or line. They are asymmetrical, meaning they have no symmetry.

Feeding Fashions

Because of their diverse forms and lifestyles, marine invertebrates have developed different ways of finding food. Below are the four main eating styles you'll find in the ocean.

Active Filtering

Animals such as sponges pump water through their body and filter out small particles of food.

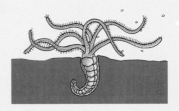

Passive Suspension Feeding

Fan worms wave hairlike cilia through the water like nets, trapping tiny bits of food.

Deposit Feeding

Sea cucumbers and some clams munch on food particles that have settled on the ocean floor.

Capturing Prey

Many marine invertebrates, such as octopuses and jellyfish, capture and kill living animals for food.

Sponges The Simplest Invertebrates

The simplest sponge looks something like an empty bag—it's hollow on the inside with a large opening at or near the top. It has no heart, liver, stomach, or any other kind of **organ.** Instead, it has different kinds of cells that help keep the animal alive. Some cells are in charge of food, others take care of reproduction, and some help the animal keep its shape. But despite the sponge's apparent simplicity there is a lot happening inside.

A sponge's soft body is supported by a kind of **skeleton** made up of hard pieces called spicules. In some sponges the spicules are made of silica—the same substance used to make glass. In others, the skeleton is soft and elastic, made of a material called spongin.

Sponges eat by filtering out food as they pump water through their body *(next page)*. A sponge the size of a pen can pump more than 20 l (5 gal.) of water through its body in a day!

Then & NOW!

Useful Sponges

For centuries, people have taken advantage of the sponge's absorbency. Ancient Greeks and Romans used natural sponges as mops and paintbrushes. Soldiers carried them into battle as lightweight drinking cups. Mothers bathed their babies with them.

Natural sponges *(top)* are really the soft spongin skeletons of dead animals. Today, they are still popular for bathing. But for household cleaning, people have copied nature by creating the synthetic sponge *(bottom)* in many different sizes and shapes.

Flat Sponges

Encrusting sponges form a thin, flat layer over rocks, shells *(below)*, or dead coral.

Generating New Generations

Round Sponges

Some sponges, like this barrel sponge, are round like a head of iceberg lettuce.

Sponges reproduce in different ways. In sexual reproduction, sperm from one sponge are released into the water *(right)* and **fertilize** eggs in other sponges. In **regeneration,** when a part of the body breaks off, it grows into a new sponge. Sponges have powerful abilities to regenerate. Even if a piece of sponge is pressed through a sieve and broken into individual cells, the cells will eventually regroup and reorganize into whole new sponges *(below)*.

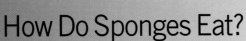

How Do Sponges Eat?

Without a mouth, a stomach, or intestines, how do sponges eat? A sponge pulls water into its body through hundreds of microscopic pores in its skin. Special cells called collar cells catch food particles like **algae** and **bacteria** from the water as it passes through. **Nutrients** from the food are picked up by another type of cell called an amebocyte, which carries the nutrients to other cells in the sponge's body. The hairlike **flagella** of the collar cells constantly wave back and forth, forcing the filtered water out through a large opening called the osculum.

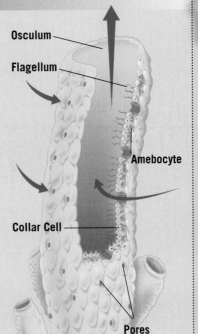

Osculum

Flagellum

Amebocyte

Collar Cell

Pores

How do we Know?

Identity Crisis

It's easy to see why the sponge was first thought to be a plant. Sponges don't appear to move at all, and many grow in branching ways. But in 1765, scientists discovered that the sponge moves enough to create water currents and for the first time called it an animal. Sponges display other animal traits as well. Unlike plants that make their own food, sponges take their food from the environment. And their cells are surrounded by cell membranes, not cell walls as plant cells are.

Tubular Sponges

Colonies of tubular sponges grow tall and straight. Water that is drawn into the sponge exits via the large hole you see at the top of each sponge.

Jellyfish and Other Stingers

A brush with a jellyfish can really hurt, because its tentacles have stinging cells that often cause allergic reactions and pain in people. These cells are called cnidocytes, and that's where the **phylum** Cnidaria gets its name. Jellyfish, sea anemones, corals *(pages 94-95),* and the Portuguese man-of-war are all cnidarians.

Cnidarians are more complex than sponges but are still pretty simple. They have a cup-shaped body with a ring of tentacles surrounding their one body opening—the mouth. Food and wastes enter and exit through the mouth. They have **tissues,** but no **organs** like a heart or stomach. And they have a network of nerve cells, but no brain.

Some cnidarians, like jellyfish, are free-floating but others, like sea anemones, live their life attached to a surface. The body of a floating cnidarian is called a **medusa,** and the attached variety is called a polyp.

Look-Alike Bodies

Cnidarians are all very different. Jellyfish drift gently through the water. Sea anemones wave tentacles that look like crazy hairdos in ocean currents. Corals live in large colonies and transfer food from one to the other.

Despite this diversity, the basic body plan of cnidarians is the same. Their body consists of a hollow pouch with tentacles around the mouth. The jellyfish *(bottom)* simply looks like an upside-down version of the sea anemone *(top).*

Methods of Multiplication

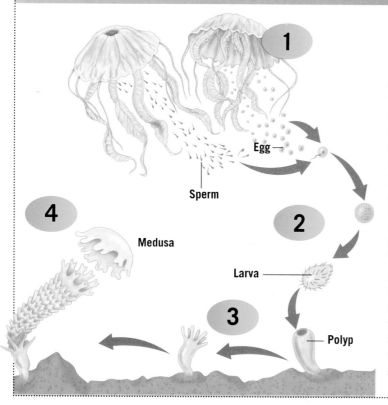

1

Egg

Sperm

4

Medusa

2

Larva

3

Polyp

Cnidarians reproduce in several ways, depending on the species. Most jellyfish reproduce sexually. Eggs and sperm are released right into the water. When a sperm joins together with an egg, the egg becomes **fertilized** *(1)* and grows into a free-swimming **larva.** The larva drifts until it finds a good place to attach itself to the ocean floor *(2).* There the larva grows into a polyp and forms buds that become tiny medusae *(3).* The medusa buds stack up, one on top of the other *(4).* Eventually, the tiny medusae break away from the polyp and grow into adult jellyfish.

Other cnidarians, like the sea anemone below, can reproduce sexually but also do so by budding. Tiny polyps stay attached to the parent until they are big enough to move away and live on their own.

Fabulous Features!

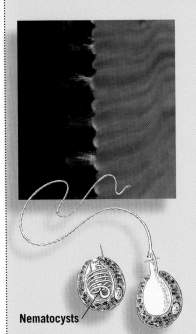

Nematocysts

Lining the tentacles of most jellyfish and other cnidarians are stinging cells, or cnidocytes. These cells contain capsules called nematocysts *(bottom, left)*. Inside each capsule rests a long thread, coiled like a spring. When triggered by touch or certain chemicals, the thread shoots out like a harpoon *(bottom, right)* and stabs the victim. Some of these threads have barbs on the end that hook into their **prey.** Others have venom that paralyzes or even kills the victim. In the photo at left, puffs of venom erupt from triggered nematocysts.

Ouch!

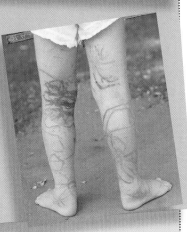

The world's deadliest creature is the box jelly, or sea wasp, which swims in the waters off the coasts of Southeast Asia and Australia. No other animal can kill a human in four minutes or less. Box jellies have a body as big as a basketball and up to 60 tentacles, each 4.5 m (15 ft.) long. The transparent box jelly is nearly invisible and doesn't attack, waiting for something to accidentally swim into its tentacles, like the prawn at top. A girl's legs *(bottom)* bear the scars of an encounter with a box jelly. Victims can survive thanks to an antivenin developed in 1970.

Chow Down

Most sea anemones stay fixed in one place throughout their life. Most eat meat. But how can they catch a meal if they can't chase it? The answer is patience.

A hungry sea anemone sweeps the water with its tentacles, waiting for a tasty morsel to come by and get a little too close *(top)*. When a fish brushes against the sea anemone's tentacles it triggers the anemone's stinging cells. The anemone injects the fish with paralyzing venom. Then it pulls the fish into its mouth with its tentacles and swallows it whole *(bottom)*.

Strange But TRUE!

The Portuguese man-of-war *(below)* looks like a jellyfish, but it isn't. It's a colony of polyps. Groups of polyps are responsible for different tasks that benefit the **colony** as a whole. Each tentacle is a polyp, and another polyp forms the air-filled bag, which catches the wind and carries the **community** along.

GIANTS of the Deep

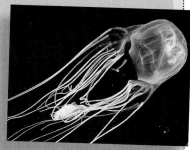

Lion's Mane

The lion's mane is the world's largest jellyfish. Some can grow to almost 2.4 m (8 ft.) across, with tentacles more than 30 m (100 ft.) long resembling a thick mane. These creatures are found as far north as the Arctic and as far south as Mexico in the Atlantic, and southern California in the Pacific.

Marine Worms
Flatworms

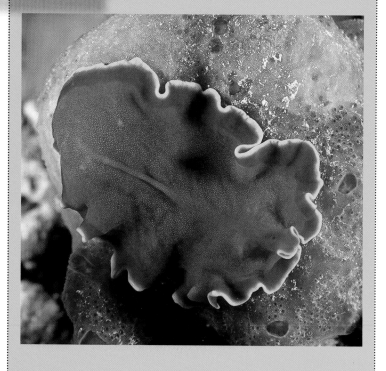

Marine worms may not look much more advanced than sponges or jellyfish, but they have an important new feature: a brain. They also are bilaterally symmetrical (just like you). This means if you drew a line down the center of their back from top to bottom, their left and right sides would look exactly the same. And they have a definite head and rear.

The soft body of most **marine** worms is supported by a kind of **skeleton** called a hydrostatic skeleton. This is a fluid-filled cavity inside their body that makes the worm rigid like a balloon filled with water. By contracting muscles that surround the skeleton, the worm changes shape and moves.

The five main types of marine worm are all named for their basic body shapes—round worm, flatworm, spoon-worm, ribbon worm, and segmented worm.

This thin, colorful creature with frilly edges may look like a pretty leaf or a petal from a flower, but it is really a worm. Because of its flat back and belly it's called a flatworm. Flatworms, the simplest of the worms, are so thin that they don't need a **circulatory system.** Instead, their wafer-thin body keeps all their cells close to the surface so they can exchange carbon dioxide for oxygen in the water.

Round Worms

Wriggling through the **sediment** of the ocean floor are huge numbers of tiny, slender worms called round worms. These worms, named for their rounded ends, are the most plentiful of all marine worms. In just 1 sq m (11 sq ft.) of bottom mud off the Dutch coast *(top)*, a scientist estimated more than four million round worms! Even in such huge numbers, we rarely notice them because most are transparent and can be seen only through a microscope *(bottom)*. Scientists have named about 12,000 species of round worms, but there may be as many as half a million more we haven't yet met.

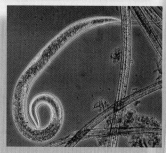

What's in a

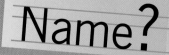

Name?

Innkeeper Worm

The fat innkeeper worm *(below)* has a reputation for hospitality. It lives in a U-shaped burrow *(left)* in ocean mud flats. By contracting and relaxing its muscles, the worm creates a current of water that brings oxygen and food into its burrow. This protected hole full of food attracts guests like tiny crabs and fish.

Fabulous Features!

When a ribbon worm senses **prey** is near, it shoots out a long, fleshy tube called a proboscis. Like a harpoon, the proboscis is often armed at the tip with a sharp dart that pierces the victim and sometimes injects it with a paralyzing venom. In some species, the proboscis is as long as the worm itself.

Ribbon Worms

How Long?

Some ribbon worms can grow to be quite long. One type called a bootlace worm is one of the longest known **invertebrates**. It can stretch to 30 m (100 ft.)—that's as long as a Boeing 737 jet airplane!

Ribbon worms have amazing powers of contraction. Some species can extend to more than 10 m (32 ft.) in length and then contract to just a few centimeters. These meat eaters feed mostly on other worms such as flatworms and round worms but also eat **crustaceans** such as crabs and shrimp, and even small fish.

Segmented Worms

The most advanced of the marine worms are the segmented worms. Their body is divided into rings, called segments. The earthworms you see in your garden are also segmented worms. These worms crawl and swim by moving each segment in sequence. They live in every part of the ocean and, as we see below, have evolved a wide range of appearances and living habits.

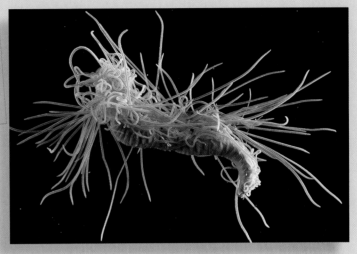

Bristle Worm
The "bristles" that run along the edges of this worm are called setae. They are like little feet or fingers that help the worm burrow, crawl, and swim.

Christmas Tree Worm
Hungry but stuck in one place, the aptly named Christmas tree worm extends spiraling tentacles into the water to filter out tiny particles of food. The worm's body is protected by a hard tube.

Fan Worm
Fan worms extend feathery tentacles into the water and trap food particles in a layer of **mucus.** When threatened, the worms retract their "fans" into tubes built out of sand and mucus.

Clams Slugs and Snails

Clams, slugs, and snails are part of the **phylum** Mollusca, the second largest phylum in the animal **kingdom.** In the ocean, there are more **species** of **mollusks** than any other **animal.**

Mollusks live everywhere in the ocean, from the shallow edges of the sea to the deep-sea floor (*pages 102-103*). Most have a soft body protected by a shell. Unlike other **invertebrates,** most have an **organ** called a radula that looks like a tongue. It is used to scrape food from rocks and other surfaces.

Clams, oysters, and mussels are **bivalves,** which means "two valves." Their soft body is tucked into a shell made up of two parts called valves.

Slugs and snails have a shell with only one valve. They are part of the biggest group of mollusks—gastropods, a word that means "stomach-footed." They got that name because it looks like they glide on their stomach.

GIANTS of the Deep

Giant Clam

The largest shelled mollusk is the giant clam found in the Indian and Pacific Oceans. It can live for more than 200 years and grow to be more than 1.2 m (4 ft.) long and close to 270 kg (600 lb.). **Algae** living in the clam's mantle give it its beautiful blue green color (*page 25*).

The Inside Story

Clams are far more complex on the inside than they look on the outside. Like us, they have organs such as a heart, a stomach, and intestines that form complex organ systems. A **circulatory system** carries **nutrients** and oxygen to the **tissues.** An excretory system expels fluids full of waste. Clams use gills as a respiratory system to absorb oxygen from the water and release carbon dioxide.

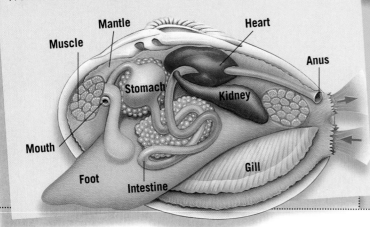

Muscle · Mantle · Heart · Stomach · Kidney · Anus · Mouth · Foot · Intestine · Gill

Making Sense

Scallops, close relatives of the clam, have dozens of eyes along the edges of their fleshy mantle. These tiny blue orbs allow them to see in all directions. Scallops also sense their surroundings with tentacles that reach out around the edges of the shell. These tentacles "smell" odors, or chemical changes, in the water.

Much in Common

The body of mollusks like clams, snails, and squid comes in many different shapes. No matter how they look on the outside, though, mollusks have three body parts in common: a foot, a visceral mass, and a mantle. The muscular foot helps the animal move from place to place. In the octopus and squid, this foot is broken into numerous parts called arms or tentacles. The visceral mass contains the internal organs. The mantle, a thin membrane covering the visceral mass, produces the mollusk's shell.

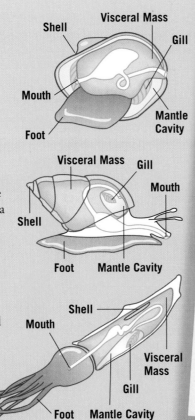

Shell
Visceral Mass
Gill
Mouth
Mantle Cavity
Foot

Visceral Mass
Gill
Mouth
Shell
Foot
Mantle Cavity

Shell
Mouth
Visceral Mass
Gill
Foot
Mantle Cavity

Slow Killer

Tropical cone shells are slow but deadly. To catch a meal, they stab passing fish, worms, and snails with poison-filled teeth on their tonguelike radula. The **prey** is paralyzed within seconds, and in some cases, like the fish below, swallowed whole.

Shell It Out

For 2,500 years, North American Indians used the shells of a rare mollusk called a *Dentalium* as money. Dentalium shells were valuable because they came from a limited area and were hard to get from the seafloor. Indians pried the animal from its long shell, which could then be used for trading. Clothing decorated with dentalium shells, like this Alaskan ceremonial headdress *(above, right),* symbolized tremendous wealth.

Where Do Shells Come From?

Mollusk shells are made of calcium carbonate, a mineral that is taken out of the water by the animal. As a clam, oyster *(below),* or other mollusk grows into an adult, the mineral is **secreted** by the mantle and hardens into a shell.

Adult oysters shed sperm and eggs into the ocean.

Fertilized eggs develop into free-swimming **larvae.** The shell starts to grow at this early stage.

The larvae eventually sink to the ocean floor, where they settle and mature.

Nudibranchs Go Naked

Believe it or not, some of the most beautiful animals in the ocean are slugs. Nudibranchs, or sea slugs, are mollusks that have no shells, leaving their soft body exposed. For protection, some are effectively camouflaged to blend in with their surroundings. Others baffle scientists with their ability to swallow the stinging cells of sea anemones, corals, and other cnidarians *(pages 38-39),* which they then adopt for their own use. The stinging cells pass through the sea slug's intestines without firing their poison and settle in the tips of the slug's fingerlike projections *(right, bottom).*

Octopuses and Squid

Octopus Anatomy

The octopus and the squid are smart. Some people think that they are the smartest **invertebrates,** maybe even smarter than some **vertebrates.** They have a complex **nervous system** directed by a well-developed brain.

These intelligent animals are part of a group of **mollusks** called cephalopods, which means "head-footed." They are called this because it looks like their feet are attached directly to a head. But what looks like a head is actually their body *(drawing at right).*

Squid and octopuses are some of the fastest swimmers in the ocean. They draw water into their mantle cavity and then force it out through an opening called a funnel. By moving the funnel, they can control their direction. A squid can swim so fast that it actually pops out of the water and flies through the air—sometimes for 30 m (100 ft.) or more before splashing back into the ocean!

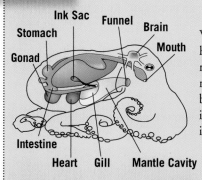

Ink Sac, Funnel, Brain, Stomach, Mouth, Gonad, Intestine, Heart, Gill, Mantle Cavity

What looks like the octopus's head is actually its entire body, minus the legs. Inside the rounded sac is the octopus's brain, along with its stomach, intestines, and all of its other internal **organs** *(left).*

Octoflage

An octopus can change its skin color to match its surroundings or to express different emotions. For instance, when frightened, an octopus turns white so it looks bigger. When angry, it turns red. It can do this because its delicate skin contains special color cells. The animal controls the size of these cells through muscle contractions. The octopus is the master of camouflage. Not only can it change its coloring to match whatever is around it, it can also adjust the texture of its skin to blend in with rocks and sand *(left).*

GIANTS of the Deep

Giant Squid

Can you imagine an invertebrate large enough to battle a sperm whale and win? Scars on sperm whales show that's just what happens sometimes when they try to eat the giant squid. These deep-sea dwellers measure up to 20 m (66 ft.) long, including their tentacles.

 Ocean Life

Crunch!

The octopus, like other cephalopods, has a pair of beaklike jaws. The hungry hunter bites through the shells of its favorite foods—crabs and lobsters—with this mighty beak. In captivity, an octopus may eat 20 to 30 crabs a day!

Relatives

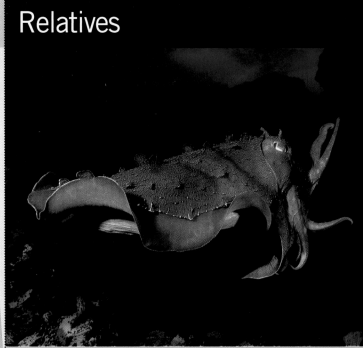

Ink Attack!

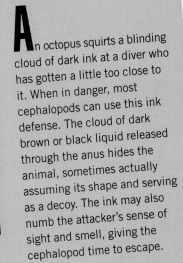

An octopus squirts a blinding cloud of dark ink at a diver who has gotten a little too close to it. When in danger, most cephalopods can use this ink defense. The cloud of dark brown or black liquid released through the anus hides the animal, sometimes actually assuming its shape and serving as a decoy. The ink may also numb the attacker's sense of sight and smell, giving the cephalopod time to escape.

Cuttlefish

Like the squid, the cuttlefish has eight arms and two longer tentacles, but its body is flattened. It also has a fin that runs along its sides. These **animals** often live on the seafloor. They have an internal shell called a cuttle-bone that helps keep the mollusk afloat. The cuttle-bone is sold as a source of calcium for pet birds. The ink sac of some cuttlefish is a source of sepia, a rich brown **pigment** used by artists.

Let's Compare

Eye Structure

The octopus has good vision. In fact, its eyes are very similar to human eyes. The octopus's pupil, though, is rectangular rather than round *(right, top)*. Another difference is in the optic nerve. Each human eye has just one optic nerve taking messages to and from the brain. The octopus has many optic nerves, resulting in a jumble of "wires" behind the eye *(right, bottom)*.

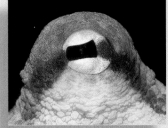

Optic Nerve   **Optic Nerves**

H u m a n   **O c t o p u s**

Nautilus

The nautilus is the most primitive cephalopod. It has existed for hundreds of millions of years and is the only one in this group that still has its outer shell.

The spiral or coiled shell is divided into many chambers *(left, bottom)*. The inner ones are filled with gas and liquid that help the nautilus control its depth in the water. The body of the nautilus lives in the outermost chamber. Attached to the body are more than 90 arms.

Octopuses Close up

Master of Disguise

The octopus doesn't have a hard internal **skeleton** the way we do. Nor does it have a rigid outer shell. It's basically a flexible bag of fluid. Because of this, the octopus can shape itself into just about anything. One kind of octopus called a mimic octopus shows how skillfully this flexibility can be used to its advantage. By assuming the shape, coloring, and movements of different sea creatures, the mimic octopus hides from potential **predators** and attracts unsuspecting **prey.** Here we see this crafty **mollusk** pretending to be *(from top to bottom)* a feather star, a baby cuttlefish, a flounder, a sea snake, and a stingray.

Motherpus

The female octopus lays thousands of eggs in a nesting den, usually a cave or a rock crevice. Unlike most other **invertebrates,** she stays and guards her eggs until they hatch *(above).* During this time, which can last up to six months, she never leaves her den, not even to eat. Newborn octopuses are tiny, only about 7 mm (0.25 in.) in length. Few survive to adulthood. An octopus can lay up to 70,000 eggs. Only about two will live to be adults. Can you find the tiny baby crawling on the eggs?

Octo-Gone!

This octopus, found off the coast of California in Monterey Bay, uses the best disguise of all. It makes itself transparent! Practically invisible, only its eyes and a few other organs can be seen.

It Takes Brains

When faced with a problem, the octopus is smart enough to find a solution. To prove this, researchers put a lobster in a clear glass jar with a cork lid and left it in front of an octopus *(1)*. Hungry for one of its favorite foods, the octopus stretched out a tentacle to snatch up the lobster but was baffled to find an invisible barrier between it and its meal. Confused, the octopus jumped on the jar and tried to bite the lobster with its beak. No luck. Three hours later, the persistent octopus finally figured out how to pop the cork top *(2)* and grab its prize *(3)*.

Small but Deadly

The Australian blue-ringed octopus only gets to be about 20 cm (8 in.) from the tip of one tentacle to the tip of another. But its bite injects a paralyzing venom potent enough to kill a person. In fact, the only two known deaths from octopus bites can be blamed on this tiny rock dweller. Both victims were dead from respiratory failure within two hours of being bitten.

I Was There!

Fred Bavendam, an underwater photographer, recalls a friendly wrestling match he had with a Pacific giant octopus off the coast of Quadra Island in British Columbia:

"With two arms it had grasped one of my bright orange strobe lights. . . . For more than five minutes we grappled. The octopus was so strong I could not pull the strobe free using force alone without risking damage to my equipment. But by repeatedly stripping away one arm and then the other, I broke the animal's hold and regained . . . possession of my equipment. Thwarted, the octopus retreated to its cave and stared out at me. All in all, I felt as if I had had a playful game of tug rope with a frisky dog."

Crabs Lobsters and Shrimp

Crabs, lobsters, and shrimp are related to the insects and spiders you see flying and crawling about on land. Like insects, they are members of the largest **phylum** in the animal **kingdom:** Arthropoda. The word means "jointed legs," a characteristic common to all the **invertebrates** in this group. Not all of these jointed legs are used for walking. Some are claws, some are jaws, and some are pincers.

An arthropod's body is protected by a hard suit of armor called an **exoskeleton.** This external shield doesn't grow with the animal, so occasionally the arthropod has to shed it, a process called **molting.** Have you ever seen a soft-shell crab? Well, that's a crab who has just molted its shell. Until the new shell hardens, the arthropod is extremely vulnerable.

In the ocean, one group of arthropods called **crustaceans** outnumber all others. They are so plentiful, they are sometimes called the insects of the sea. Lobsters, shrimp, and crabs are crustaceans. These animals have a well-developed **nervous system.** They smell and taste food through two pairs of antennae, which are covered with thousands of sensory nerves. They can find food 20 m (66 ft.) away in the dark!

GIANTS of the Deep

Giant Spider Crab

The largest crustacean in the world is the giant spider crab, found only in a small area off the coast of Japan. Its body grows to almost 45 cm (18 in.) long and 30 cm (12 in.) wide. But what really makes this crab look big is the span of its legs, which can measure 4 m (13 ft.) from side to side.

Express Yourself

Crabs are very expressive creatures. They use body postures and leg and antenna movements to convey such emotions as love, fear, and anger. The crab below has just been approached by a potential enemy. First, it puts its claws in front of itself as if to say: "You better watch out." It then rises up on its legs and waves its claws as if to say: "If you get any closer, I'll attack you!" Next, it crouches down low into a defensive position. Finally, the crab turns its body and prepares for a quick retreat.

CAUTION

ATTACK

DEFENSE

RETREAT

Surprise! It's a Shrimp

Can you see the shrimp in this picture? The photographer who took the picture thought he was taking a picture of a kind of coral called a bubble coral. But when he looked closely through his lens he was surprised to discover a tiny, almost invisible shrimp—only 7 mm (0.25 in.) long—resting on the coral!

Periscope Eyes

The eyes of many crabs and lobsters are at the end of moving stalks called peduncles that act like little periscopes. The surface of each eye is rounded, giving the crustaceans a wide field of view, usually 180 degrees or more. Although their view is good, what they actually see is much fuzzier and less distinct than what we see.

Better Than Floss

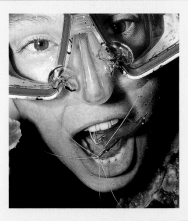

Looking for a new dentist? Just jump in the ocean and open wide. That's what this diver did. Noticing a cleaner shrimp picking away parasites and food bits from the mouth of a large fish, she opened her mouth and the crustacean dentist got right to work. No appointment necessary.

Lobster Line Dance

How **Far?**

When ocean temperatures cool in the fall, the American spiny lobster heads south to warmer waters. But the migrating arthropod usually doesn't travel alone. Thousands of lobsters travel together, forming lines of up to 65 marching single file. Traveling day and night they can cover 16 km (10 mi.) in 24 hours.

Relatives

Barnacle

For a long time, barnacles were classified as **mollusks.** It's easy to see why. From the outside, they look a lot like clams and mussels. But they are crustaceans. Adult barnacles are attached to solid surfaces and live upside down, extending their legs up and outside their thick protective plates to filter food from the water.

Horseshoe Crab

The horseshoe crab is not really a crab at all. It is an arthropod, but it is more closely related to spiders. Horseshoe crabs are called living fossils because they are almost identical to their ancestors who lived 250 million years ago. Their long, spiny tail and armored body make them look threatening. But they are harmless to humans. In fact, they are actually quite helpful. Scientists have learned a lot about how our eyes work by studying the horseshoe crab's eyes, which have some of the largest light receptors in the **animal** kingdom.

Sea Stars Spiny Slowpokes

A Stomach Turner

Sea stars don't move very fast, about 15 cm (6 in.) per minute, but they never have to turn. That's because they don't have a head or a tail end, so when they walk they can lead with any arm.

Sea stars walk on tube feet, special structures that are part of an elaborate water-filled system of vessels. Throughout the sea star's body are interconnected canals that carry water from the center of the animal to each of its arms. The water canals transport food and wastes while carrying oxygen and carbon dioxide to and from the body **tissues,** much like the blood in our body.

Sea stars and their relatives were the first **animals** to develop an internal **skeleton,** more than 650 million years ago. The hard plates of the skeleton often poke out, making the animal's surface bumpy. That's why these animals are called echinoderms, which means "spiny skinned."

Sea stars don't bring food to their stomach, they bring their stomach to the food. When a sea star finds a clam, for example, it sits on the shell and pulls on it with its tube feet. Eventually, the clam's muscles get tired and its shell opens a tiny bit. That's enough for the sea star, who quickly pushes its stomach into the clam's shell. The clam's flesh is turned into a liquid that the sea star "drinks" through its stomach. When it's done—up to 15 hours later—the sea star pulls its stomach back inside its body.

Lost and Found

A lost arm or two is no cause for alarm if you're a sea star. You just grow new ones *(below).* If a big enough chunk of the sea star's center breaks off with the arm, it can grow into a whole new sea star! This process is called **regeneration.** But regeneration is not always a quick fix. It can take up to a year for a starfish to completely regrow a lost part.

All Shapes and Sizes

Sea stars, also called starfish, are not always drab yellow with five arms. Many species dress up in red, orange, purple, green, and other bright colors. Some are even two-toned.

The arms of sea stars come in a variety of lengths and numbers. Some have long, slender arms *(bottom),* whereas others have stumpy ones, making them look more like biscuits *(center)* than stars. Typically, sea stars have five arms, but this varies, too. If they have more than five arms they are often in multiples of five. Some species have as many as 50!

Fabulous Features!

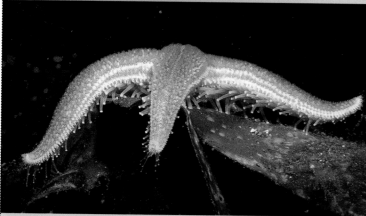

Lining the underside of sea stars are hundreds of tiny, finger-like projections called tube feet. These help the sea star walk and grip **prey.** Each tube foot is connected to a little sac of water. When the sea star contracts muscles around the sac, water is squeezed into the tube foot, making it expand like a water balloon. The foot is then solid enough for the sea star to push against as it walks or grasps prey. The movements of the tube feet are coordinated so they push and pull together. The tube feet can also grip surfaces by suction, allowing the sea star to hold on tightly when it's being pounded by waves.

What's inside Those Arms?

The round center of the sea star houses its stomach. Its mouth is below the stomach, on the bottom of the sea star, and its anus is above the stomach. Near the anus is the vascular pore, where water enters the sea star to fill its hundreds of tube feet. Each arm contains the rest of the sea star's internal **organs.** Skeletal plates give the animal internal support and protection. Digestive glands absorb the **nutrients** from food. Sexual organs, or **gonads,** contain eggs or sperm for reproduction.

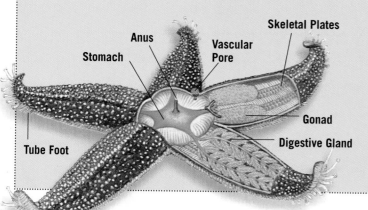

Anus
Stomach
Vascular Pore
Skeletal Plates
Gonad
Digestive Gland
Tube Foot

Relatives

Sand Dollar

Beachcombers commonly find the round, flat skeletons of sand dollars, bleached white by the sun *(inset).* When sand dollars are alive *(left),* they live buried in the soft and sandy parts of the ocean floor. They sift through the ocean bottom with their tube feet, searching for small food particles. Like all echinoderms, they are covered with spines, but theirs are very short.

Sea Cucumber

Another bottom-dwelling echinoderm, the slow-moving sea cucumber, has developed some unusual defense mechanisms. Several **species secrete** toxic substances when disturbed. Still others have a more extreme response to danger. They suddenly eject their internal organs through their mouth or anus. The action may startle and confuse the attacker enough for the sea cucumber to escape. Soon it will grow new internal organs.

Sea Urchin

Of all the echinoderms, the sea urchin is the spiniest. The round body is covered with spines—some as long as 30 cm (12 in.). Because of these threatening spikes, many animals avoid sea urchins. But people of the past have found uses for some of these spines. The spines of the slate-pencil sea urchin, for example, were once used as pencils for slate boards.

Sea Squirts — Salps and Lancelets

Sea squirts are potato-shaped **marine** invertebrates also called tunicates because their saclike body is protected by a leathery or gelatinous outer covering called a tunic.

Sea squirts live their adult life stuck in one place, but they start out as free-swimming **larvae.** They paddle around for a few days until they find a good place to settle down, usually the surface of a rock, and there they attach themselves for life. Some sea squirts live alone *(small photo at right),* but others live in **colonies** jumbled loosely together *(large photo at right).*

Like you, sea squirts are part of the **phylum** Chordata. We all share certain characteristics at stages in our development, including a flexible rod of **cartilage** down our back. This rod, called a notochord, is replaced by a backbone in humans and other **vertebrates.** But in sea squirts, as well as their relatives the salps and lancelets, no backbone develops. That's why they are called **chordates** without a backbone.

Water Pumps

Most sea squirts have two openings called **siphons.** Water enters through the siphon on the top and exits through the one on the side.

More Than Meets the Eye

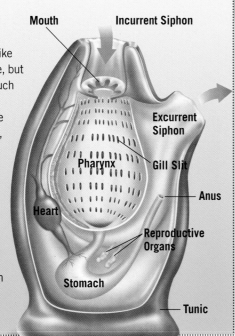

Sea squirts may look like sponges from the outside, but on the inside they are much more complex. Unlike sponges, sea squirts have **organs,** including a heart, stomach, intestines, and reproductive organs. A basketlike filter called a pharynx strains food and oxygen from water that enters through an opening at the top (incurrent siphon) and exits through an opening on the side (excurrent siphon).

Labels: Mouth · Incurrent Siphon · Excurrent Siphon · Gill Slit · Pharynx · Heart · Anus · Reproductive Organs · Stomach · Tunic

Salps

Closely related to sea squirts are the transparent salps. Like sea squirts, salps are filter feeders, but they take in water through an opening at one end and force it out an opening at the other end, like a jet engine. This water current helps them move. Salps live alone or in colonies that can form long floating chains that are several meters long *(right).*

Our Relatives

It may be hard to believe, but sea squirts are some of our closest invertebrate relatives. To see what we have in common, you have to look at the sea squirt before it is grown up. Developing sea squirts look like tadpoles *(below)*. At this stage, they have the four features common to all chordates, including humans. They have gill slits, a **dorsal** nerve cord, a notochord, and a muscular tail. But as they grow, the notochord and tail disappear and a filtering sac and siphons develop from the gill slits.

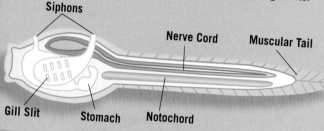

Siphons

Nerve Cord

Muscular Tail

Gill Slit

Stomach

Notochord

What's in a

Name?

Sea Squirt

Normally, a sea squirt draws water in through one opening and expels it out another. But if the animal is touched or otherwise disturbed, it contracts violently and squirts water out both openings *(below)*. This is how it got its common name "sea squirt."

Fishy Lancelets

Buried in the sand or fine gravel of warm, shallow ocean waters is a small animal that looks like a fish. But looks can be deceiving. This **invertebrate**, not more than 7 cm (3 in.) long, is called a lancelet because its flat body is shaped like a blade or lance. Descended from very similar creatures that lived 570 million years ago, the lancelet is thought to be the link between sea squirts and the most primitive fishlike vertebrates. The lancelet has all the characteristics of a vertebrate except a backbone. It even has the beginnings of a fin that runs along most of its body.

Medicines from the Sea

Have you ever wondered where medicines come from? Scientists produce most of them from plants, **animals,** and minerals found on the land around us. But now, in their search for new medicines, scientists are diving into the ocean.

The ocean is full of creatures—often **invertebrates**—that appear to be defenseless: sponges that can't move, soft-bodied sea squirts, fragile-looking sea whips. How do these vulnerable animals keep **predators** from munching on them? The answer is chemical warfare. Some animals manufacture substances that make them taste so awful nothing wants to eat them. Others produce deadly poisons. Predators know to stay away. Scientists are turning many of these defensive chemicals into medicines for people.

Marine animals are making a difference in the world of medicine in other ways, too. Substances from mussels and crabs are changing the ways surgeons close wounds, and pharmaceutical companies use horseshoe crab blood to test their products for **bacteria.**

No Swell

The stationary sea whip, a kind of soft coral, can't swim away when danger approaches. Instead, it produces a chemical that makes it taste gross. The chemical may disgust fish, but scientists think it's swell. A compound from this substance soothes swelling and inflammation in human skin. Someday it might be the best treatment for problems like sunburn and arthritis.

Stop Pain with Poison

When a cone shell attacks a fish *(right)*, it uses tiny "harpoons" to inject venom into its **prey.** The poison is made of as many as 30 different **proteins** that act together to paralyze the victim. Some cone shell **species** produce venom strong enough to kill people. But scientists believe that if they separate the proteins in the venom, the proteins may heal, not harm. In fact, the cone shell's venom could be the source of 30 new drugs! One such drug is currently being tested as a nonnarcotic painkiller.

Would **You** *Believe?*

A Chemical Powerhouse

Sponges may be the simplest animals in the ocean, but they produce some of the greatest chemical weapons. That's why scientists are studying them so carefully. At the right dosages, poisons from sponges can be lifesaving drugs for people. For instance, a chemical from this deep-sea sponge *(right)* may help fight cancer. Other sponge species make chemicals that relieve inflammation and swelling and that help the body accept **organ** transplants.

Before any fluids, such as intravenous solutions, can be injected into your bloodstream, they must first be tested for bacteria. Drops of the horseshoe crab's bacteria-sensitive blue blood are added to the fluid *(left)*. If bacteria are present, the fluid curdles, or gets lumpy. The more bacteria, the lumpier it gets.

Invertebrate Cancer Killer

A sea squirt species *(below)* that grows in the mangrove swamps of the Caribbean may help in the battle against cancer. Scientists have discovered a chemical in this marine invertebrate that is being studied for its ability to fight leukemia in mice and breast cancer in humans. More than 30 other drugs derived from creatures in the ocean are being tested as possible cancer fighters. The other creature seen in the photo below is a kind of sea slug.

Super Glue from the Sea

Mussels *(top)*, oysters, and a few other kinds of **bivalves** know how to stick around. They make a super-sticky substance called byssus that glues them to a hard surface, even in the water. Scientists use a form of this super-strong glue to fix parts of human eyes and teeth, which are also wet places. In the future, the glue may even become a replacement for surgical stitches.

For now, though, doctors still use stitches to close wounds. But some doctors are threading their needles with crab shells, which contain chitin. Chitin can be turned into a natural thread that's long-lasting and won't cause infections. Chitin can also be used to make special bandages that patch wounds inside a person's body.

What Is a Fish?

More than 13,000 different kinds of fish swim in the sea, but all have some things in common. They are **vertebrates,** meaning they have a spine, or backbone. Instead of lungs, fish use gills to absorb oxygen. Most fish have a body that is covered with scales, are **cold-blooded**—the temperature of their blood matches the temperature of their surroundings—and lay eggs.

Fish have a lightweight, flexible **skeleton** and strong muscles to help them swim. Most have fins on their back, belly, sides, and tail, which help them turn, move forward, and stay upright.

Some fish live in fresh water, but the species described in this book live in salt water and are called **marine,** or ocean, fish. Not all ocean creatures with fins and a backbone are fish. Whales and dolphins breathe through lungs like we do and give birth to live young, which makes them **mammals.**

Let's Compare

Saltwater and Freshwater Fish

Freshwater fish don't need to drink water, but saltwater fish do. That's because the water inside an ocean fish's body is less salty than the surrounding seawater. A process called osmosis causes less salty water to move to salty water, so the water inside an ocean fish is constantly leaking out through its skin and gills. The fish must keep drinking to replace it. A freshwater fish has the opposite problem. Water is continuously soaking into its body, and the fish must constantly urinate to get rid of it. Saltwater fish seldom urinate.

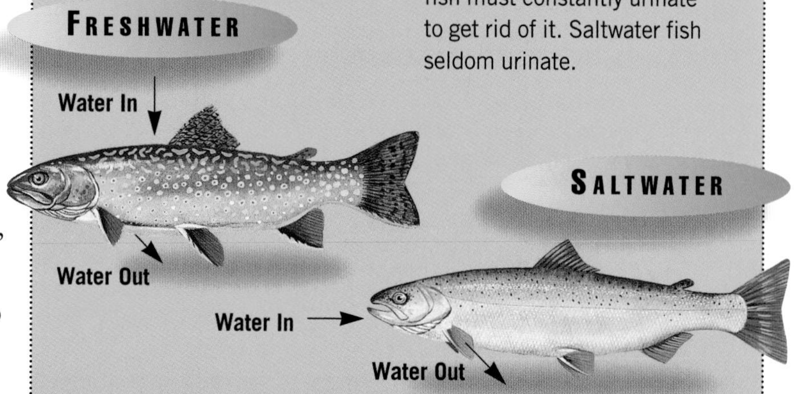

FRESHWATER

Water In

Water Out

SALTWATER

Water In

Water Out

Fish Anatomy

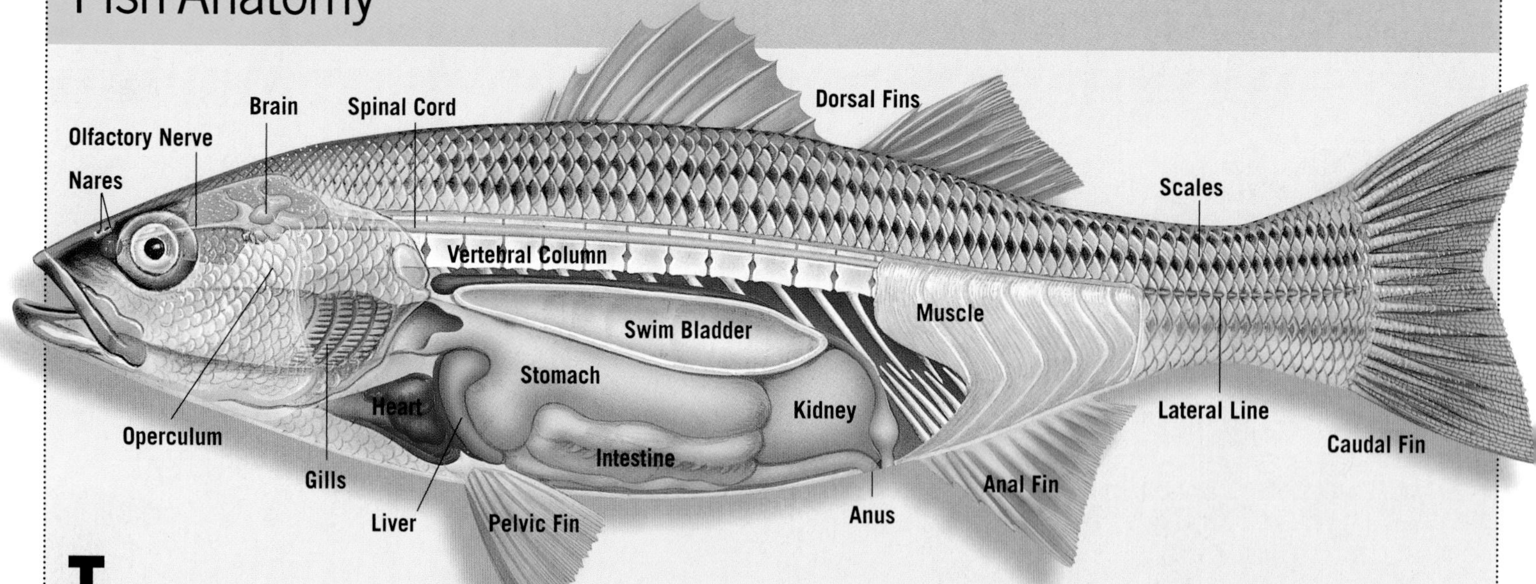

Olfactory Nerve

Brain

Spinal Cord

Dorsal Fins

Nares

Scales

Vertebral Column

Muscle

Swim Bladder

Stomach

Kidney

Lateral Line

Heart

Operculum

Intestine

Caudal Fin

Gills

Anal Fin

Liver

Pelvic Fin

Anus

The inside of a fish doesn't look that different from the inside of a human. There's a backbone, brain, heart, kidneys, stomach, intestine, and liver. In the typical fish shown above, most of the **organs** are crammed into the lower front area of the body. Swimming muscles (*pink*) fill the rest of the space. Most fish have a swim bladder that works like an air-filled balloon to keep them from sinking. The gills are protected by a covering called an operculum.

Classes of Fish

Jawless

Jawless **species** like the hag-fish *(page 59)* and lamprey *(left)* are the simplest class of fish. Their sucker mouth is lined with teeth, and they lack both a jaw and a complex digestive system.

Cartilaginous

This class of fish includes sharks, rays, and skates. Instead of bone, their skeleton is made of tough **cartilage**. Many have rows of replaceable teeth set in jaws on the underside of their body.

Bony

The largest class of fish have a skeleton made of bone. Bony fish have jaws that meet at the midline of their body, and their teeth are fixed in their jaws. Flaps cover their gills.

Feeding Strategies

Different species of fish eat in different ways. The lamprey clamps its mouth onto a fish and sucks its blood. Parrotfish have a beaklike mouth to scrape **algae** off coral. And the huge manta ray filters tiny **plankton** out of the water. Below are two more examples.

On the Prowl

The big, sharp teeth of the Atlantic wolffish point inward to prevent **prey** from escaping. Strong jaws and teeth crush the hard shells of the wolffish's favorite foods—mussels, sea urchins, and starfish—which it swallows, shell and all.

Fishing for Fish

A frogfish attracts its prey with a flap of skin that dangles off its forehead like bait attached to a fishing pole. When a curious fish moves close to investigate, the frogfish quickly opens its mouth and pulls the prey in.

Gulping Air

The green moray eel at left may be preparing to attack, or it may just be taking a big gulp of oxygen. To breathe, fish take water into their mouth and pump it over their gills. The gills act like the surface of our lungs, but in this case they collect oxygen from water instead of air. Located on both sides of the head just behind the mouth, gills are bony arches holding many gill filaments, which look like the teeth of a comb. The filaments are full of blood vessels, which absorb oxygen from the water.

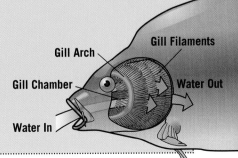

Gill Arch
Gill Filaments
Gill Chamber
Water Out
Water In

Fish Body Shapes

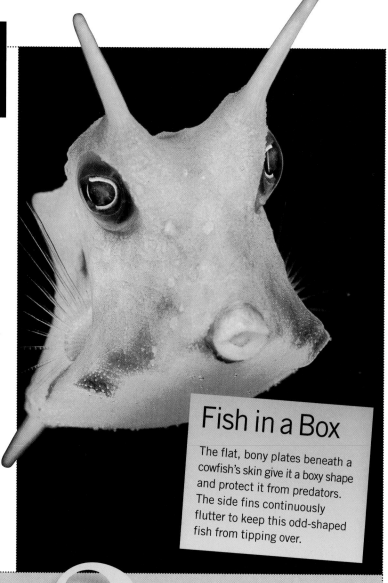

The shape of a fish can often tell you where or how it lives. For example, a streamlined body belongs to fast swimmers that roam the oceans, whereas a long, snakelike body is good for burrowing into mud or wiggling through coral.

Speedy swordfish and marlin have a smooth body and a crescent-shaped tail attached by a narrow joint called a peduncle. This joint helps the fish whip its tail back and forth while keeping the rest of its body still, so it can swim long distances without getting tired. Other fish don't travel far to find food. They lie on the ocean floor waiting for **prey** to come to them. Some of these bottom dwellers have a flat body that helps them blend into the seabed.

Fish with a narrow, disk-shaped body can maneuver easily through underwater vegetation. They can change direction quickly to escape, and their thin shape makes them hard to see when facing a **predator** head-on.

Fish in a Box

The flat, bony plates beneath a cowfish's skin give it a boxy shape and protect it from predators. The side fins continuously flutter to keep this odd-shaped fish from tipping over.

Would You Believe?

Sideways Fish

A flatfish, like the peacock flounder at left, starts its life as a normal-looking fish. At birth, it swims upright, with one eye on each side of its head (far right). As it grows it becomes very flat, and one eye gradually moves to join the other eye on the opposite side. A flatfish spends the rest of its life lying on its blind side on the ocean floor. Some flatfish have both their eyes on the left side of their body, whereas other **species** have their eyes on the right side.

Torpedo

The barracuda *(below)* is shaped like a torpedo, with a narrow head, wider middle, and tapering tail. This shape is common to many fish, especially predators and long-distance swimmers. It helps them cut through the water at top speed.

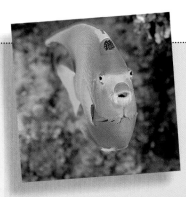

Flat

Flat as a pancake standing on its side, the queen angelfish *(above)* is shaped to fit its environment. A thin, disk-shaped body, about as wide as the rim of a dinner plate, allows it to weave through forests of underwater plants and among the nooks and crannies of coral reefs.

Round

This Hawaiian guinea fowl pufferfish *(below)* isn't always round. It's just a trick used to scare off predators. Like porcupinefish and other pufferfish, this species can inflate its body like a balloon by swallowing large amounts of water.

Tubular

The hagfish *(above)* can swim right side up or upside down by contracting the muscles in its long, tubelike body. This shape is well suited to living in burrows on the muddy ocean floor. When feeding, the hagfish sometimes knots its body into a pretzel shape to hold its prey.

Tail of a Sea Horse

What fish has a head like a horse, a snout like an anteater, and a tail like a monkey? It's the sea horse! Its tail is called prehensile because it can grab things—the tail of another sea horse, for example, as these two are demonstrating in a mating dance. If they want to travel, sea horses will often hitch a ride by curling their tail around a blade of drifting sea grass. To stay still, they hook onto a piece of coral.

How Heavy?

Ocean Sunfish

Its German name means "swimming head," and that's just what the ocean sunfish looks like. Weighing up to 1.8 t (2 tn.), it is the heaviest bony fish in the sea. Whale sharks weigh more, but they're cartilaginous, not bony, fish.

How Do Fish Move?

Fish usually get around the ocean by swimming, but not all fish swim the same way. Some **species** are good at quick acceleration, like a sprinter that can go from a standing start to top speed in seconds. Other fish can switch directions quickly, like a basketball or soccer player. And some fish can swim at a strong steady pace over long distances, like a marathon runner.

Most fish swim with their whole body, but others rely more on their fins. The tuna flicks its powerful tail to shoot forward. The sea horse flaps the small **dorsal** fin on its back to swim. The electric eel moves by waving the anal fin on the underside of its body. Flounders ripple their dorsal and anal fins to move along the ocean floor, and they can also lift their body straight up by squirting water through their downward-facing gill.

How Fish Swim

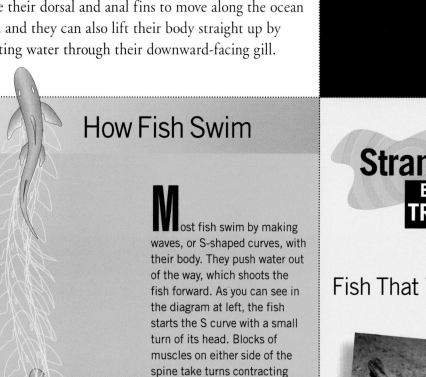

Most fish swim by making waves, or S-shaped curves, with their body. They push water out of the way, which shoots the fish forward. As you can see in the diagram at left, the fish starts the S curve with a small turn of its head. Blocks of muscles on either side of the spine take turns contracting and relaxing. This motion turns the fish first in one direction, then the other, the way a snake wriggles on land. The S curve continues down the body, getting larger and stronger until it reaches the tail.

Strange But TRUE!

Fish That Walk

The batfish lives on the bottom of the Pacific Ocean near South America. It hardly looks or moves like a fish at all. This strange creature can prop itself up on its pelvic and **pectoral fins,** which it then uses like arms and legs to scoot rather than swim along the ocean floor.

Predator and Prey

The great barracuda, which can grow to almost 2 m (6 ft.) long and weigh 45 kg (100 lb.), is an awesome **predator**. Its long, slim body and pointed head cut through the water at high speed. The fins clustered together at its rear end like the feathers on an arrow help it shoot forward for a quick attack. The young crevalle jacks chasing this barracuda away are also strong, fast swimmers. Their short, disk-shaped body helps them to maneuver and suddenly change direction.

A Gallery of Tails

The rounded tail of the porcupinefish helps push its body forward. This tail is also good for steering between rocks and for keeping the fish upright.

This fan-shaped tail belongs to a grouper, which lies in wait for **prey** to swim by. A tail like this helps a fish accelerate quickly to catch a meal.

With a deeply forked tail shaped like the wings of a jet, a cruising fish like the yellowtail snapper can build up great speed. This tail, however, is not designed for quick acceleration.

The sea bream has a triangular tail, good for accelerating, stopping, and sudden turns, all of which are helpful to a fish that roams the ocean bottom looking for food.

A Garden of Eels

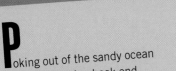

Poking out of the sandy ocean floor and swaying back and forth with the current, garden eels look more like sea grasses than animals. They dig holes with their strong tail and then plant themselves, waiting for **plankton** to drift by. When predators or storms threaten, they duck inside the holes. Thousands of eels live together in "gardens" that can cover an area as large as a football field.

Inspired by the streamlined shapes of fish, 1950s automobile designers often added two large tail fins to the rear end of a car, like the Cadillac at right, to give a sense of speed.

Fish Senses

Fish experience their world in the same way that we humans experience ours—through the five senses. They can see, hear, taste, touch, and smell. Fish even have a sixth sense that we don't have.

But fish senses don't work exactly like human senses do. Some fish, for example, have taste buds on their lips and other body surfaces. Most fish rely more on their sense of smell to find food than on their eyesight. Sharks are well known for their ability to smell blood a long way off. When a shark first notices a scent, it will swim back and forth, trying to pinpoint which direction the scent is coming from. When it finally determines the location—from up to 0.8 km (0.5 mi.) away—the shark will lock onto that course until it finds its **prey.**

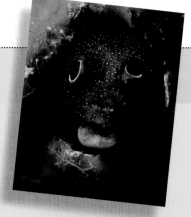

The muscles in each eye of a fish work independently, so one eye can look up while the other looks down *(above).*

Hearing

Fish don't have ears that stick out of their head, but they do have an inner ear that resembles the inside of a human ear. Inside a fish's ear is a liquid-filled capsule containing three semi-circular canals and three small bones called otoliths. The canals help balance the fish. The otoliths are surrounded by hairs that pick up sound waves from the water and then send signals to the brain. Some fish have additional hearing parts called the Weberian apparatus. These are a set of small bones that connect the swim bladder to the inner ear. They help the fish hear a wider range of sounds.

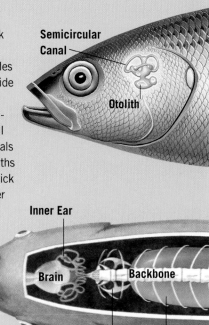

Semicircular Canal

Otolith

Inner Ear

Brain   Backbone

Weberian Apparatus   Swim Bladder

The Eyes Have It

Fish eyes come in many colors and varieties, as you can see in the examples below. Some of these patterns help a fish camouflage or hide itself. Other designs help fish see better by controlling the amount of light that reaches their eyes. Fish don't have eyelids, but the requiem shark *(below, left)* has a slit pupil to reduce the amount of light its eyes receive, and the ray *(below, right)* uses a built-in eyeshade to protect its eyes against sudden changes in brightness.

Wrasse

Requiem Shark

Sea Bream

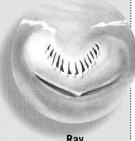

Ray

The Sixth Sense Knows

Fish have a sixth sense that lets them "feel" even the smallest movements or vibrations in water created by currents or other creatures. They do this by using an **organ** called the lateral line—a system of tiny canals running along the sides of their body under the skin.

On some fish, like the yellow seaperch *(below)*, you can see it clearly. Inside are tiny hairs connected to nerves that send messages to the brain. Lateral lines help schools of fish move and turn in perfect formation and also prevent aquarium fish from smashing into the invisible walls of their tank.

The Touchy, Feely Sea Robin

The sea robin has three separate spines on each **pectoral fin** that look like the legs of a spider. It uses these spines to walk along the bottom and to feel around for food. Even more unusual, the spines have taste buds, so the sea robin can taste things with its "fingers."

Strange But TRUE!

Fish don't have vocal cords like we do, but they can still make sounds to communicate with each other. They grunt, moan, hiss, croak, knock, or whistle, often using their teeth or bones to make these noises. The drumfish at right is one of the ocean's loudest fish. It vibrates muscles against its **swim bladder** to make a drumming sound that carries long distances underwater. One reason for the drumming is to locate a mate.

Drumming Up a Companion

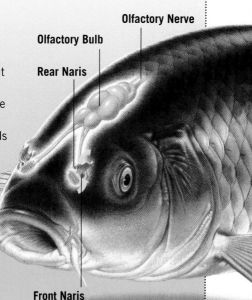

Sense of Smell

Fish use their sense of smell to find food, avoid enemies, and identify members of the opposite sex. On their snout they have holes called nares that work sort of like our nostrils. Water enters the front naris, bringing in odors, and exits the rear naris. Inside, the smelling organ, or **olfactory** bulb, is lined with folded walls covered with nerves that absorb smells and send information about these smells to the olfactory center of the brain. The olfactory center can identify smells and even tastes.

Olfactory Nerve

Olfactory Bulb

Rear Naris

Front Naris

Fish Reproduction

When it's time for fish to lay eggs, they usually travel to the same spot every year. They find a place where the temperature and other conditions are just right for baby fish to hatch. In a process called spawning, the mother fish sheds thousands of eggs into the water, where the father **fertilizes** them.

Some ocean fish, like the salmon, shed their eggs in fresh water, traveling hundreds of miles every year to their favorite spawning grounds. The grunion actually lays eggs on land, burying them on the sandy shores of California beaches, where they hatch and return to the ocean. Some sharks don't shed eggs but give birth to live young *(page 69)*.

Most mother fish leave their eggs to hatch and grow on their own. Many of the baby fish, called fry, become food for other fish before they reach full size. But some fish parents, like the species described at right and on the next page, protect their eggs and fry.

Egg Watchers

Although most fish don't guard their eggs, the smooth frogfish is an exception. The mother and father take turns watching over their developing **embryos,** or unborn babies, which can be seen in the golden mass of transparent eggs in the photo above. When a fish comes looking for an easy egg meal, the parent eats the would-be predator.

Would **You** *Believe?*

Gender Switch

When is a female angelfish not a female angelfish? When it turns into a male angelfish! Several **species** of tropical fish like the angelfish have the ability to function as either males or females. Such fish are called hermaphrodites. When a male angelfish living with a group of females dies, the largest and oldest female changes into a male and takes his place, fertilizing the eggs and protecting the group from **predators**.

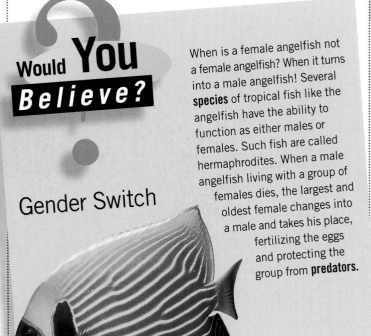

How Many?

Eggs

Because so many eggs and fry become food for other fish, mother fish release an enormous number of eggs to guarantee that at least some survive. The champion is the ocean sunfish *(right),* which sheds 30 million eggs at each spawning. If every one hatched and survived, the sea would be packed with ocean sunfish from one shore to another.

Parenthood

Leafy Sea Dragon

It may look like a moving clump of seaweed, but the creature at right is a leafy sea dragon, a relative of the sea horse found only off Australia. Sea dragon parents like this one protect their babies by giving them piggyback rides.

Cardinalfish

Most species of cardinalfish are mouthbreeders, which means they carry their eggs in their mouth. Once the female lays the eggs, the male gobbles them up into his mouth. When they are ready to hatch, he spits them out—very carefully, though, so he doesn't pierce them with his sharp teeth.

Strange But TRUE!

Fantastic Fathers

A pregnant father? It's true! Sea horses give birth to live young, but instead of the mother delivering the babies, it's the father! The mother lays her eggs in a special pouch on the front of the father's body, where the eggs are fertilized and start to grow. She visits him every day for the next three weeks. The father's body delivers food and oxygen to the developing babies. As the babies grow, his pouch gets bigger and bigger. When the babies are ready to be born, the father contracts the muscles of his pouch, and out pop the tiny sea horses.

Jawfish

Like the cardinalfish above, the jawfish father also uses his mouth as an egg hatchery. He holds the eggs safely in his mouth for one or two weeks until the babies are ready to be born.

Fish Defenses

It's a dangerous ocean out there, especially if you're a fish. To increase their chances of surviving to adulthood, many fish have ingenious ways of protecting themselves. Some use defensive techniques, such as hiding themselves against the ocean floor through camouflage coloring. Other fish use more straightforward means of defending themselves, such as prickly spines or lethal poisons to scare away **predators.**

Fish that swim in large groups, called schools, such as the ones at right have a distinct advantage; some oceanographers think that sheer numbers confuse the enemies of these fish. Faced with thousands of possible **prey,** the theory goes, all swimming together in a hypnotic parade, some predators can't make up their mind which ones to eat and end up going away hungry. Many **species** of fish that depend on schools for protection have shiny, silvery sides, and that only adds to the enemy's confusion.

Fooling the Enemy

Can you guess which one of these fish is the deadly moray eel, and which one is the harmless look-alike? You'd be smart to do what most predators do; rather than make a dangerous mistake, they avoid both. The plesiopid fish in the bottom photo is harmless —its only defenses are the polka dots and the phony eye on its tail, which make it look like the moray's head. At the first sign of danger, the plesiopid darts headfirst into a hole, leaving its tail exposed.

Buddy System

Sometimes a fish will team up with another sea creature for security. The black-rayed goby and the snapping shrimp, for example, have worked out a relationship that benefits both of them. The shrimp doesn't see very well so it shares its home with the sharp-sighted fish. The shrimp also catches food for them both. In return for a home and meals, the goby warns the shrimp when danger approaches.

Try to Find Me

Where can you hide wearing brilliant red or orange? In a coral reef, you would fit right in with the brightly colored environment. Fish in the open ocean blend in better when colored silvery green, blue, or brown. Some fish have a shape that helps them disappear into the background. See if you can spot the camouflaged fish below.

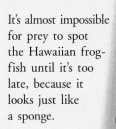

These two speckled sand dabs perfectly match the pebbly texture and the color of the sea bottom where they live.

It's almost impossible for prey to spot the Hawaiian frogfish until it's too late, because it looks just like a sponge.

The painted frogfish at left matches the yellow sponge garden it lives in. It can also change color in a few short minutes.

This ghost pipefish stands out against a dark background but disappears among the green plants where it makes its home.

Pufferfish

To keep from being eaten by predators, the pufferfish makes itself too big to swallow. You can see the normal size and shape of the masked pufferfish in the top photo at left. When it senses danger, the pufferfish gulps a lot of water and inflates itself to about two or three times its normal size *(bottom photo)*. As an extra defense, the pufferfish is poisonous.

Poisonous Fish

Don't step on a stonefish! The needle-sharp spines on its **dorsal** fin are filled with poison that can kill a human being in 20 minutes. Unlike other poisonous fish whose bright patterns warn other fish away, the stonefish lies camouflaged on the ocean floor. Can you see it?

Yikes! Spikes!

En garde! These fish defend themselves with spikes that protrude in all directions. The beautiful scorpionfish *(top)* has sharp, poisonous spines on all of its fins except the tail. The balloonfish *(bottom)*, a type of porcupinefish, raises the spiky scales on its body when it is alarmed.

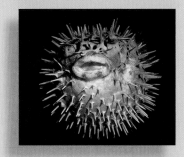

Sharks

Sharks have been around for more than 300 million years, since before the dinosaurs. During that time, they have evolved into extremely efficient eating machines. All sharks are **carnivores,** which means they eat animals; they even eat other sharks!

They differ from bony fish in several ways. Their **skeleton** is made of **cartilage,** not bone. They don't have **swim bladders** to keep them afloat, so they must either swim or sink. Their gills are not covered with a protective flap of skin. Most shark babies are born alive, rather than hatching from eggs. And sharks have an amazingly good sense of smell.

Not all **species** of sharks are ferocious killers, and most are not dangerous to humans. In fact, sharks can be very beneficial. Their cartilage is used to treat cancer, their liver is rich in vitamin C, and their eyes have been used to help people who need cornea transplants.

Strange But TRUE!

Covered in Teeth

If you rub a shark the wrong way, you'll cut your hand! As you can see in the photo of the swell shark below, a shark's skin is covered in tiny, razor-sharp points, which are more like teeth than scales. They cover the shark from snout to tail, providing protection and helping water run off the shark's body.

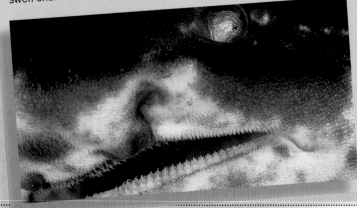

GIANTS of the Deep

Whale Shark

The world's largest living fish, the whale shark, is a harmless giant. The divers swimming alongside it *(above)* give you some idea of its size—14 m (45 ft.) long and up to 18 t (20 tn.). With its 2-m (6-ft.)-wide mouth, the whale shark scoops up loads of tiny **plankton** and small fish such as anchovies and sardines.

Fabulous Features!

Electrical Sensors

Unlike bony fish, sharks and rays use electricity to help them find their **prey.** These electrical sensors, called the ampullae of Lorenzini after the scientist who discovered them 300 years ago, are located in small, deep pits in the skin on their head. They can pick up the weak electrical signals emitted by other creatures. The hammer-head shark *(below)* has these pits across its extra-wide snout, which may give hammerheads an extra advantage in detecting electricity.

Canal Ampullae of Lorenzini
Pits
Skin

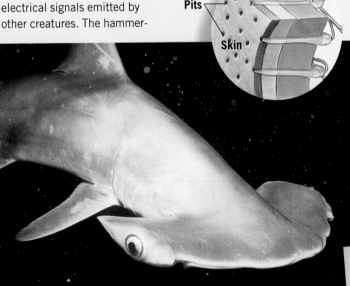

Shark Teeth

Sharks never stop growing teeth. New rows of teeth move forward on a continuous belt as old teeth wear out. You can see these rows of teeth clearly in the shark's mouth at right. Young lemon sharks may go through a set of teeth in a week. Shark teeth come in a variety of shapes and sizes, depending on what prey the shark eats. You can tell which species of shark made a bite by looking at the teeth marks.

Shortfin Mako **Blue Shark** **Horned Shark** **Great White Shark**

A Mouth for Every Meal

Sand Tiger

When you think of a shark's mouth, you probably picture the one in the top photo at right belonging to a sand tiger. Its mouthful of narrow, jagged, and needle-sharp teeth are perfectly designed for seizing and tearing flesh from prey.

But not all sharks eat the same way. The basking shark in the middle photo swims with its giant mouth open, to catch as many tiny plankton as possible. The mouth of the horned shark *(bottom photo)* is shaped like a pig's snout and has small, pointed front teeth for grabbing and holding prey. The strong, flat teeth at the back of the mouth are used for crushing and grinding sea urchins and shellfish.

Basking Shark

Horned Shark

Birth of a Shark

They're called pups when they're born, but don't be fooled by the cute name. These baby sharks are dangerous! Unlike most fish, which hatch from eggs, sharks are usually born alive *(upper photo)*. Some, however, begin to grow as eggs inside the mother, then finish developing outside her body in egg cases attached to seaweed *(lower photo)*, living off **nutrients** in the yolk sac. The firstborn, or hatched, sometimes eats the unborn brothers and sisters, and any unfertilized eggs.

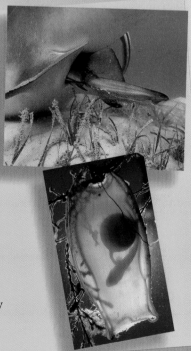

Sharks Close up

I Was There!

In the clear water, the sun made spidery patterns along the shark's back as it slowly circled. I leaned through the cage's camera port. The shark approached. I backed into the cage, but my tank caught on the camera port's upper bar. I was trapped. I pushed the shark away with my camera housing . . . and managed to pull my head inside.

On the next pass, the shark bit tentatively at the cage's float . . . and looked at me. Its eye was a soulless lens, not a creature's eye. Pushing its snout at the camera port, it could not get its lower jaw inside. It opened its jaws and banged its snout against the bars. I looked down its throat as its bottom teeth scraped the dome of my camera housing.

The shark stayed with us the entire afternoon. . . . Two smaller sharks approached. With the day's last light, the wind came up, the sea's surface became corrugated, and the sharks vanished.

—Underwater photographer David Doubilet, from his book Light in the Sea

Underwater photographer David Doubilet wasn't scared when he took this picture, not until "I put down the camera and realized how close the teeth were."

Safe in a steel cage, shark researcher Dr. Eugenie Clark watches great white sharks cruise past, attracted by bait called chum.

Sleeping Sharks

People used to think sharks would die if they stopped swimming and pumping water over their gills, but these dochizame sharks lying in a cave off the coast of Japan prove that's not true. They are barely moving and appear to be taking a nap. Shark researcher Eugenie Clark discovered that the cave water contains more oxygen than normal, which may explain why the sharks in here don't die.

Shark Gallery

Although the familiar shark has a streamlined, cigar-shaped body and grows 2 to 2.15 m (6 to 7 ft.) long, there are many that don't fit this description. The hammerheads shown below are perhaps the strangest-looking sharks. Their head is shaped like the letter T, with one eye and one nostril at each end. Unlike most sharks, they frequently travel in schools.

Horned Shark

The California horned shark has a pair of hornlike spines on its back, one in front of each **dorsal** fin. Its eyes are set in high ridges, and dark spots cover its body.

Wobbegong

This tasseled wobbegong looks more like a carpet than a shark. Its wide, flat body is covered with fleshy flaps that make it difficult to see against the ocean floor.

Devil Shark

One of the world's smallest sharks is found in Suruga Bay off the coast of Japan. It is shown fully grown in this picture.

Then & NOW!

prehistoric jaws
life-size model from ancestor of great white shark

If you think today's sharks look scary, you should have seen them 16 million years ago! The giant *Carcharodon megalodon*, whose jaw replica appears at right, was similar to the great white shark of modern times, except that it was twice as big. You can compare the size of its fossilized tooth with the size of a great white's tooth in the photo above. This monster was the largest carnivorous shark of all time, measuring about 13 m (40 ft.) long and weighing 23 t (25 tn.).

Rays and Skates

Sharks aren't the only cartilaginous fish in the sea—they also have cousins, the rays. If you took a shark and flattened it, you'd get something that looks like a ray. Skates are one of the **species** of ray.

Unlike sharks, which swim by moving their tail back and forth, skates and rays seem to fly or sail through the water, flapping the tips of their **pectoral fins,** or chest fins, up and down. Rays live on the ocean floor. Their eyes are perched on top of their body instead of on the sides, which gives them a good view of what's around and above them, but not what's below.

On the underside of their body are their gills and mouth. This can make breathing messy for a fish that lives on a muddy bottom. But the ray sucks in clean water through holes, called spiracles, on the top of its head just behind the eyes, then squirts the water out through its gills on the bottom.

Eating Styles

This southern stingray opens its mouth to accept a ballyhoo from underwater photographer and diving guide Jay Ireland. He discovered that stingrays would gather regularly at a spot where fishermen cleaned their catch off Grand Cayman Island in the Caribbean. Ireland began feeding the stingrays, who learned to expect food from humans and became pushy when divers did not offer food.

Let's Compare

Different Types of Rays

The seven families of ray include stingray, manta ray, electric ray, eagle ray, sawfish, guitarfish, and skate. Rays, like the blue-spotted ribbontail stingray *(top),* have a tail like a whip, with a sharp, often poisonous barb in the middle for defense. Skates *(bottom)* have a divided pelvic fin and a thicker tail, often with two small **dorsal** fins and a caudal fin at the end. Most rays bear live young, but skates lay eggs.

Guitarfish

The guitarfish is named after the musical instrument it resembles. It has the flattened head and body of a ray, but a shark's dorsal fins and tail. The guitarfish hasn't changed much in 35 million years, and may be the oldest type of ray.

Social Animals

Rays have the largest brain and are the most sociable of cartilaginous fish; at least they don't eat each other, as sharks do. Some kinds of rays live alone, but others, such as the bat ray *(below)*, travel long distances in large groups. Some groups of rays include hundreds or even thousands of individuals. Rays are known to gather in favorite areas, where, if space is limited, they lie on top of one another. They appear to enjoy being stroked on their sensitive skin. Like sharks, they have ampullae of Lorenzini *(page 69)*, which can detect other animals. They use these sensors to spot **predators, prey,** and possible mates.

GIANTS of the Deep

Manta Ray

The largest ray of all, the manta, measures 7 m (22 ft.) across—almost the wingspan of a small plane! The huge fins sticking out of the front of its head give manta rays the nickname devilfish. But mantas are not dangerous to humans; they feed only on **plankton,** shrimp, and small fish.

Hiding Out

Can you spot the ray? With eyes on the top of its head, the ray can hide its body completely under mud or sand and still keep a lookout for predators or prey. Above its eyes are frilly flaps of skin covering breathing holes called spiracles. The ray is so good at camouflaging itself that divers sometimes step on one by accident.

Sea Turtles

Like their land relatives, sea turtles live inside hardened shells. Only they spend most of their time in the water, sticking their head out now and then to breathe. Sea turtles have very good vision, smell, and hearing. They can zip through the water at 29 km/h (18 mph)—more than three times faster than the fastest human swimmers. There are eight **species** of sea turtles—Australian flatback, black, green, hawksbill, leatherback, loggerhead, Kemp's ridley, and olive ridley. The smallest are the ridleys, which don't get much bigger than 75 cm (30 in.) in length. Leatherbacks are the largest, averaging 1.8 m (6 ft.) long.

Where in the World?

Sea turtles are found throughout the Atlantic, Pacific, and Indian Oceans.

Let's Compare

Sea Turtles and Land Turtles

Over the millions of years they have been on earth, sea turtles *(top)* have adapted to life in the water. Their very sleek, flat shape allows them to move easily through the water. Their claws have fused into paddlelike flippers, and they have powerful forelimbs for swimming. They cannot, however, retreat into their shell like their cousins the land turtles, often called tortoises *(bottom)*.

Against All Odds

When sea turtles are ready to lay eggs, they emerge from the water and dig deep holes on a sandy beach *(above)*. They lay their eggs inside and cover them with sand. Sometimes the eggs are found and eaten by birds, raccoons, and other animals. Even humans are a threat. Turtle eggs are a delicacy in some cultures. If undisturbed, the eggs will hatch about two months later. The tiny baby turtles push their way to the surface and rush to the sea *(below)*. They rush for a good reason—on land they are vulnerable to attack from **predators**. With such dangers, only one out of every 10 baby turtles makes it to the sea. Of these, only one or two in a thousand survive to adulthood.

Green Turtle Migration

The green turtle is a really good swimmer. An adult can swim underwater for up to five hours before coming up for air! One group swims 2,250 km (1,400 mi.), from the coast of Brazil to Ascension Island in the middle of the Atlantic Ocean, to mate and lay eggs. The trip takes nearly a month. Two months later, they turn around and return to

Brazil
Ascension Island

Brazil. When the babies hatch they, too, make the long swim to Brazil.

Tortoiseshell

For centuries, tortoiseshell has been prized for its great beauty and durability. It has been crafted into hair combs, cases, and even eyeglass frames.

Because of its richly patterned shell, the hawksbill turtle *(right)* of the Atlantic Ocean was hunted nearly to extinction and is now on the endangered species list. Fortunately, today plastics are used to create most "tortoise-shell" accessories.

In the 18th century, tortoiseshell was a popular material for making combs and eyeglass frames.

GIANTS of the Deep
Leatherback Turtles

The leatherback turtle *(below)* has a tough, leatherlike covering, not a hard, bony shell like other sea turtles. It is the largest turtle in the world—on land and in the sea. It can grow up to 180 cm (6 ft.) long and weigh 0.9 t (1 tn.).

People — The Marcovaldis

Sea turtles have been rapidly disappearing from the world's oceans. Their greatest enemy has been man, who has hunted them for their skin to make leather, their shell to make accessories, and their meat and eggs for food. Sometimes they drown when they mistakenly get caught in fishing nets.

But conservationists are trying to stop that. In Brazil, for example, husband-and-wife oceanographers Guy and Neca Marcovaldi *(above, right)* run the National Sea Turtle Conservation Program (or TAMAR). Started in 1980 with government funding, TAMAR employs 400 people and runs 21 patrol stations, covering more than 990 km (620 mi.) of coastline. TAMAR's efforts have paid off—in 19 years, they have rescued more than 2,000 adult sea turtles and helped 2.8 million hatchlings survive.

Although it is now illegal to trap sea turtles, they still attract a lot of attention when they come to shore. Tourists flock to the beaches to photograph these remarkable creatures.

Sea Snakes

Silent Hunter

A sea snake glides effortlessly and silently through the azure waters around a coral reef off the coast of Australia. Hidden in the reef's nooks and crannies are the eels and other fish that may become the snake's next meal! Sea snakes range in size from 60 cm (2 ft.) up to 1.8 m (6 ft.).

Sea snakes spend most if not all of their life at sea. There are about 50 **species,** more than any other **marine** reptile.

Although sea snakes look like the snakes you see on land, they have some unique features that make it easier for them to live in the water. They have a flattened body that helps them move through water. They have nostrils on top of their head, so they can take in oxygen just by floating up to the surface. They are able to stay underwater for hours with one breath. Many give birth in the water, so they do not risk their life by going up on land like sea turtles do *(page 74).*

Sea snakes are beautiful but deadly—they inject venom, a poisonous liquid, with their bite. Fortunately, they are not aggressive and rarely attack humans.

Where in the World?

Sea snakes are found in the Pacific and Indian Oceans and the Caribbean Sea.

Fabulous Features!

One of the neatest things about sea snakes is their tail—it is flattened and broad, like the paddle of a rowboat. And just like a paddle, this feature helps the **reptiles** propel themselves through water and steer themselves in the direction they wish to go.

Jormungand

There are many ancient legends about sea snakes. One Norse myth tells of the terrible Jormungand, one of the three monstrous children of the god Loki and the ogress Angerboda. Odin, head of the Norse gods, knew that these creatures would cause trouble, so he threw one into the land of the dead and tied up the second with a magic ribbon. Jormungand the snake was thrown into the sea, where he grew and grew. He grew so big and long that he encircled the world, his fangs biting into his own tail.

Would YOU Believe?

Deadly Venom

All sea snakes have venom, injected through the fangs into their **prey** to subdue them. The beaked sea snake, found in the South Pacific, has the deadliest venom—one single drop could kill five human adults.

Yellow-Bellied Sea Snake

The yellow-bellied snake is the most widespread sea snake. It is found in the Indian and Pacific Oceans, from the east coast of Africa to the west coast of Central America. Its distinctive colors make it a most striking snake. It has a black stripe running along its top, a bright yellow underside, and a diamond pattern on its tail. It is poisonous, but only a fourth as deadly as the beaked sea snake (above).

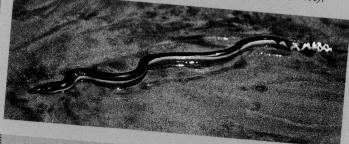

Strange But TRUE!

Sea of Snakes

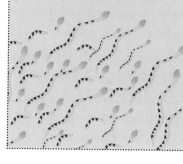

Sea snakes sometimes travel along the surface of the water in a large group called a snake slick. In 1932 in the Strait of Malacca—a body of water between the Malay Peninsula and the island of Sumatra—a man witnessed an enormous slick that was 3 m (10 ft.) wide and 100 km (60 mi.) long!

One reason so many snakes travel together may be that sea snakes are surface feeders and find it easier to feed and reproduce when there are a lot of them around.

Sea Kraits

Although it is a type of sea snake, the sea krait is much fonder of land than other sea snakes are. The sea krait likes to crawl ashore now and then and rest on rocks. It also likes to prowl uninhabited islands. Unlike other sea snakes that give birth in the water, the sea krait lays its eggs on land, usually in caves and crevices in rocks. Sea kraits are brightly colored snakes, encircled by dark bands of color.

Sea Otters

Some of the most cheerful-looking creatures are sea otters, who live in the coastal waters of the Pacific Ocean. They are the smallest of **marine** mammals, although they can be the size of a child—about 1.2 m (4 ft.) long and 14 to 45 kg (30 to 100 lb.).

Sea otters are a lot like land **mammals.** They have external ears, well-developed forepaws, and short but distinct back legs. They also have a thick coat of fur, which is waterproof and thus helps them keep warm. But take one look at their big flippers and you know these cute, furry creatures are meant to live in the water!

When not diving through the water in search of food, sea otters are usually floating on their back. They eat, sleep, groom, and look after their young this way—the mother putting her pup, or baby, on her chest.

Tools of the Trade

Very few animals use tools like we humans do. Chimpanzees do, some birds do, and all sea otters do. With their forepaws, they use rocks to break shellfish away from ledges. They also use rocks to break open the hard shell of such food as clams and sea urchins they bring to the water's surface *(above, left)*. While floating on their back, they open the shells of their dinner by pounding them against a rock lying on their stomach *(above)*.

Would **You** *Believe?*

A Sea Otter's Hairy Hair

Mammals are **warm-blooded,** so for them to live in the sea requires special protection from cold and moisture. The extreme thickness of sea otter fur keeps them warm—6.5 sq cm (1 sq in.) may contain nearly a million hairs, 10 times the number of hairs we have on our entire head! When an otter cleans itself, it pushes air into the fine underfur, which provides insulation and increases its ability to float.

Safety in Mother's Arms

The baby sea otter is helpless, so the mother has to keep a watchful eye all the time. While she floats on her back, the baby sleeps or plays on her stomach. When swimming through the water, she holds him in her forearms. Sometimes, when she needs to hunt for food, she ties him loosely in a mass of seaweed to keep him in place until she returns.

Oh, Kelp, Sweet Kelp!

Sea otters tend to stay near beds of sea kelp, a seaweed with thick leaflike blades. The beds are part of a vast network stretching to the ocean floor. Kelp beds *(pages 26-27)* are like an underwater forest, providing shelter and food for a large **community** of marine **animals,** including fish and shellfish.

Living in the kelp beds is very convenient for sea otters. There's plenty of food, and at night they wrap kelp around themselves to keep from drifting away from home.

Saving Pups from Harm

Sometimes baby sea otters are separated from their mothers. This is serious, since the pups do not know how to feed or clean themselves. California's Monterey Bay Aquarium has set up the Sea Otter Research and Conservation (SORAC) program to rescue orphans.

First, human caretakers feed young pups a special formula of liquefied seafood, vitamins, and minerals. They groom their fur so they can stay dry and warm. After three months, the babies are fed solid foods such as clams and squid *(left)*. Then the pups attend "sea otter school," where they learn skills they'll need to survive in their natural **habitat.** At six to seven months, the sea otters graduate and are set free into the open ocean.

Capturing an Otter's Spirit

Native American artists skillfully depict animals in drawings and sculptures. These are often done in a stylized, rather than strictly realistic, way. That is because the artwork is meant to capture the spirit of the creatures, not their exact likeness, as in this lively sculpture by Kwakwaka'wakw artist Henry Hunt. It shows a sea otter in a typical pose—floating on its back, a sea urchin on its stomach, ready to be cracked open.

Seals and Walruses

With their sleek body and strong flippers, seals, sea lions, and walruses slip quickly and gracefully through their watery world. Their flippers look like fins, so these **animals** are known as **pinnipeds,** which means "fin-footed."

Pinnipeds can swim long distances but they are as comfortable sunning themselves on land as they are swimming through the water. While in the water they must breathe air, so they stick their head out of the water every now and then. They can, however, stay underwater for up to 20 minutes at a time and are able to dive down several hundred meters in search of food. They eat fish, squid, shrimp, crabs, and other shellfish.

In size they range from the Galápagos fur seal—the female is about 1.2 m (4 ft.) long and weighs about 30 kg (60 lb.)—to the southern elephant seal, which can weigh more than 100 times that.

Pinnipeds are generally found in oceans, but some are found in lakes and landlocked seas. Walruses live in herds, whereas seals travel alone or in a small group.

Catching Rays

Even from a distance, walruses can easily be identified by their two long white tusks. These are extensions of their front teeth and can be up to 1 m (3 ft.) long. They are very useful. With them the walrus can pull itself up on land, break a hole in the ice for breathing, or use them as weapons in fighting.

When walruses lie in the sun, their skin turns pink as the blood rushes to the surface of their skin. This helps their blood to warm up.

Let's Compare

Seals and Sea Lions

Sea Lion

Seal

What's the difference between seals and sea lions? Sea lions *(top)* have flaps over their ears, and their back flippers can bend forward, which helps them get around on land.

Seals *(bottom)* don't have ear flaps, and their hind flippers can only stay behind them, which forces them to wriggle like giant slugs on land. In the water, sea lions use their front flippers to swim *(circle, top)*, whereas seals mainly use their back flippers *(circle, bottom)*. The one exception to this is fur seals—like sea lions, they have ear flaps and can "walk" on all four flippers.

Wonderful Whiskers

Those stiff whiskers you see on seals, sea lions, and walruses are not just for looks. They are very sensitive and help these pinnipeds sense the movement of fish and other creatures in the water around them. They can also help in the search for food along the ocean floor. A walrus uses its bristly whiskers like fingers to feel for food in mud and sand. It can be dark on the ocean bottom, so pinnipeds often have to depend on their sense of touch rather than their eyesight when searching for food.

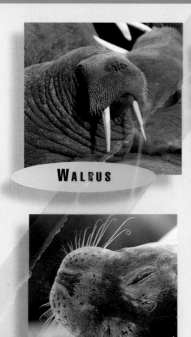

WALRUS

SEA LION

GIANTS of the Deep

Southern Elephant Seal

The largest pinniped of all is the southern elephant seal, which can reach a length of 6 m (20 ft.) and can weigh almost 3.6 t (4 tn.)—that's heavier than two cars!

This seal is named for the male's trunklike nose. At mating time, the male fills his snout with air so that it becomes extended and blows out a loud bellow to scare off rivals.

Mane Event

Sometimes animals get their names because they look like other animals. This is the case with sea lions—the adult male sea lion sometimes has a mass of hair surrounding its face resembling the mane of an African lion.

They can also be rather ferocious and look like a roaring lion *(above)*. Baby sea lions try to stay out of their way to avoid getting stepped upon or injured.

Deadly Net

Commercial fishing nets, up to 40 km (25 mi.) long, pose a real danger to animals living in the sea. Every year thousands of **marine** animals—including pinnipeds, dolphins, porpoises, turtles, and birds—are killed when they are accidentally trapped by these nets. They become so entangled that they cannot move and eventually drown. Fortunately, the sea lion in this photograph was rescued by human beings and set free from its prison.

Bobbing Asleep

The ocean is one big waterbed for sea lions. They often sleep while bobbing up and down in shallow water. Once asleep, they slowly sink to the bottom, then float to the surface to breathe before sinking down again. In deeper water, they float vertically near the surface, with their nostrils poking just above the water line so they can breathe.

Manatees

The manatee also goes by the name sea cow because, like a land cow, it is big and bulky and feeds on vegetation. But a manatee is more closely related to an elephant—it is descended from the same ancestor—and looks like a cross between a sea lion and a whale. It has a small, blunt head and tiny eyes. To move about, it uses two flippers located on either side of a fat torso that tapers into a paddlelike tail.

The average manatee is 2.5 to 4.5 m (8 to 14 ft.) long and weighs 200 to 600 kg (440 to 1,300 lb.). To achieve that size, they eat a lot—10 to 15 percent of their body weight per day! Slow and gentle, they graze on plants growing along riverbeds and the floor of coastal regions.

Manatees are found in only a few places—Florida's waterways and coastal waters, along the Gulf of Mexico, in the Caribbean Sea, and along the west coast of Africa. They, and their relative the dugong, prefer warm tropical and subtropical climates.

What's in a Name?

Sirenia

Manatees and their relative the dugong were once mistaken for mermaids, the legendary creatures with the head and torso of a woman and the tail of a fish. Mermaids were also known as sirens, thus the Latin name for these sea creatures is Sirenia.

At certain angles from afar, manatees do look somewhat human. And when they nurse their young, they resemble a woman cradling a child.

Steller's Sea Cow

Manatee Conservation

The population of manatees has been dwindling. Today in North America there are only about 2,000 manatees, living mainly along the coast of Florida. As an endangered species, they are protected by law. They have no natural enemies, but they are threatened by people disturbing them in their **habitat** —particularly people in speedboats. Manatees swim close to the water's surface *(right)*, and because they are so slow and sluggish, it is hard for them to avoid oncoming boats. Injuries from boat propellers are so common that scientists now use scars *(top, right)* as a way to identify individual manatees.

Manatee Greeting

The manatee's way of saying "hi" is to go up and nuzzle its friend. They gently touch noses and chirp and squeal, perhaps having a conversation just like you do when you see a friend.

Some creatures are not with us anymore due to overhunting. One of them is Steller's sea cow *(left, bottom)*, a large sirenian that was discovered in the Bering Sea by Russian explorers in 1741. It was about twice as long as a manatee *(left, top)*. Unfortunately, once these big, gentle animals were discovered they were hunted so relentlessly that they were extinct within 30 years.

Relatives

Dugong

Although manatees live in both fresh water and seawater, their relative the dugong lives only in salt water. They also look slightly different. Manatees have a paddlelike tail, whereas dugongs have a notched tail, like a dolphin. Dugongs go by the nickname sea pigs because they like to use their snout to root around in the seabed.

Dugongs are found only in the temperate waters of the Indian and Pacific Oceans, mostly along the coasts of East Africa, Asia, and Australia. About 85,000 live off the coast of Australia.

Meet the Whale Family

Whales are the giants of the sea. The blue whale is the largest animal that ever lived on earth. They are bigger than the dinosaurs that existed in prehistoric times. Fortunately, water helps support their enormous size, and their sleek, torpedo-shaped body moves easily through their liquid environment.

There are two kinds of whales—toothed and baleen. Toothed whales have teeth and include the smaller members of the whale family, dolphins and porpoises. Baleen whales have a **baleen,** which is a stiff filter that hangs down from the upper jaw of the mouth. Despite being so huge, baleen whales are gentle creatures and feed on tiny organisms like krill, a shrimplike creature, and **plankton** *(pages* 28-31*).*

Whales are **mammals,** like us. Their scientific name is Cetacea. There are 71 species of toothed whales and 10 species of baleen whales.

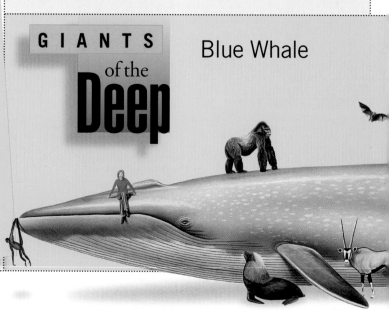

Whales have nostrils on the top of their head! These nostrils are called blowholes—toothed whales have one and baleen whales have two. Underwater, the blowhole shuts tight to keep water out. When the whale surfaces, the skin around the blowhole immediately opens up. Then the whale breathes in and out very quickly. The steamy white "spout" we see shooting above its head is the blowing out of air mixed with water droplets at high pressure.

What's Echolocation?

By sending out sound signals —a kind of clicking noise—and listening to the echoes that return, dolphins and whales can figure out the size and distance of objects ahead. This sonar system, called echolocation, helps them to find food and navigate underwater.

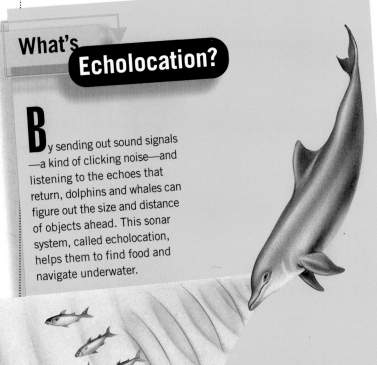

GIANTS of the Deep

Blue Whale

Whale Gallery

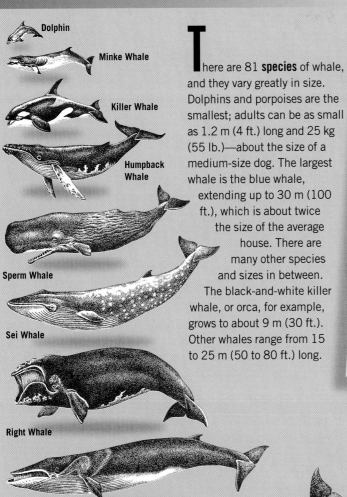

Dolphin

Minke Whale

Killer Whale

Humpback Whale

Sperm Whale

Sei Whale

Right Whale

Fin Whale

Blue Whale

There are 81 **species** of whale, and they vary greatly in size. Dolphins and porpoises are the smallest; adults can be as small as 1.2 m (4 ft.) long and 25 kg (55 lb.)—about the size of a medium-size dog. The largest whale is the blue whale, extending up to 30 m (100 ft.), which is about twice the size of the average house. There are many other species and sizes in between. The black-and-white killer whale, or orca, for example, grows to about 9 m (30 ft.). Other whales range from 15 to 25 m (50 to 80 ft.) long.

The blue whale is the giant of the giants, weighing more than 150,000 kg (330,000 lb.). Just the tongue of an adult blue whale weighs 3,000 kg (6,500 lb.), about the weight of two cars! At birth, a baby blue whale is already more than 6 m (20 ft.) long —greater than three times the height of an adult man! When fully grown, these whales dwarf all other living creatures.

Helping Out

Most whales are communal animals. They live in groups called pods and help one another in many ways.

Female dolphins work together to protect their young. During long-distance swims they keep a close watch over the young dolphins to make sure they don't stray. When they aren't traveling, the mothers circle around the babies, forming a protective "playpen" *(right)*.

Sometimes whales and dolphins help push their injured members up to the surface to breathe. And of course they work together as a pack to hunt down **prey**.

Stranded

Once in a while dolphins and other whales get stranded on beaches. Scientists are not sure why this happens. Some say their sonar system does not work properly in shallow areas. Others blame high **tides** and lightning storms. Illness caused by pollutants or microorganisms could also be a reason.

These heavy creatures are not designed to move on land, so once they're beached they cannot get back in the water. They can die from sunburn or dehydration. Large whales may even suffocate—their tremendous weight squishes their lungs and other **organs** when their body is not supported by water.

Dolphins and Porpoises

The smallest members of the whale family are dolphins and porpoises. They are small toothed whales that can be found in the coastal waters of all the oceans. Dolphins are often seen near beaches flipping up into the air as they zip through the water.

Like all whales, dolphins and porpoises swim by pushing their tail up and down and steer by moving their flippers. They have highly sensitive hearing, which helps them communicate with one another as well as perceive what is going on around them.

Dolphins are very intelligent creatures and have been part of ocean folklore since early seafaring days. Ancient Greek and Roman legends tell of dolphins befriending people. Even today stories of dolphins helping human swimmers are common. However, despite their popular image as the friendly guardians of the sea, dolphins are wild animals. They should be observed with caution and respect.

Dolphins and porpoises like to swim in groups, called pods or herds. These may be as small as two to five members for harbor porpoises, and occasionally as big as 100,000 for deep-sea dolphins. The bottle-nosed dolphin swims in pods of five to 10, usually made up of several adult females and their young, called calves. Small groups of males swim separately. There's safety in numbers. They look after one another, and whenever one sees danger it warns the others. To hunt for fish, they may split into smaller groups.

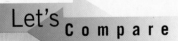

Dolphins and Porpoises

Porpoises are the smaller cousins of dolphins. Porpoises have a blunt snout and flat, triangular teeth. Dolphins have a longer, pointier snout—called a beak—and their little upturned mouth looks like a cheerful smile. They have sharp, cone-shaped teeth.

Although porpoises tend to be shy creatures, dolphins are just the opposite—they are friendly and outgoing and love to play with anything and anyone passing by. Dolphins are the great entertainers that put on shows at zoos and aquariums.

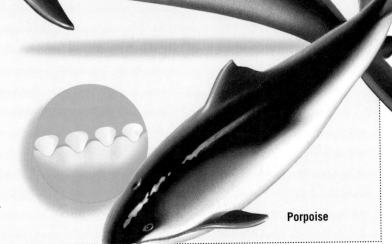

Dolphin

Porpoise

Let's Play!

Dolphins are the clowns of the ocean. They are friendly and curious, playing freely with objects or other **mammals** they meet. Sometimes they approach swimmers and ships to check them out. Just for fun, they can leap 5 m (16 ft.) into the air and "walk" on their tail.

Porpoises, on the other hand, are shy and less social—rarely jumping out of the water and preferring to live in small groups.

The Legend of Arion

Since ancient times, dolphins have been famous for their friendliness and intelligence. An ancient Greek myth tells us of Arion, who won a music contest in Sicily. On his way home to Corinth, greedy sailors robbed him of his prize and made him jump overboard. Before he did so, he sang a beautiful song, which attracted dolphins around the ship.

So when he plunged into the water, a friendly dolphin offered Arion a ride and carried him back home.

Would **You** *Believe?*

Dolphins in the Navy

Submarines and dolphins have a lot in common—they both move through water and use sound waves, or sonar, to find objects underwater. Only dolphins do both things better, and the U.S. Navy knows it. So the navy has been studying dolphins to learn from them so it can build better submarines and sonar equipment. The navy also relies on these smart sea mammals as able underwater assistants, using them to search for victims of boating accidents and to locate and mark sunken objects *(below)*.

Whales

Scientists believe that whales are the descendants of land **mammals** that went back to the sea more than 50 million years ago. These mammals were four-legged creatures similar to wolves and bears. Over millions of years, their body adapted completely to their life in the water. They lost their fur and external ears. Their body became smooth and sleek, so that water could flow easily around it. Their back legs became small and weak from lack of use, eventually evolving into a tail that spread out into two wide wings called flukes. Their arms evolved into flippers jutting out from either side of the body. If you were to look at these flippers through an x-ray machine, you would see that the bones supporting them look a lot like the fingers in your hands! Like human hands, they have wrist bones, finger bones, and joints.

Let's Compare

Teeth vs. Baleen

The toothed whales—such as killer whales, sperm whales, dolphins, and porpoises—have rows of teeth on the upper and lower jaws of their mouth. They eat fish, squid, and even larger animals found in the sea.

Hanging from a baleen whale's upper jaw is a stiff comblike filter made of the same material as your fingernails. This filter, called a **baleen,** strains tiny organisms from the water. Every so often, the whale passes its tongue over the baleen and swallows the food it finds trapped there.

Would You Believe?

The Tail Names the Whale

A whale's tail is made up of two lobes called flukes. No two pairs of flukes are alike. In the North Atlantic, the tails of more than 2,000 humpback whales were photographed and carefully studied. Each was found to have its own pattern of markings, scars, and coloring. Thus, a whale's flukes are like our fingerprints and can be used to provide positive identification for a specific animal.

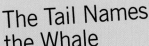
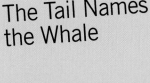

The Journey of the Gray Whale

Many whales return to the same place to mate and to give birth to their young. The California gray whales hold the distance record for this **migration.** In fact, they migrate farther than any other mammal. They swim 16,000 km (10,000 mi.) each way, a journey that takes two to three months (below). In the fall they leave their feeding grounds in the Bering and Chukchi Seas and head for the warm waters around Baja California, Mexico. Here they mate and have their babies. In the spring they go back up the coast of North America to feed.

During these long journeys, whales hardly eat at all. They live on the stores of energy in their blubber *(next page)*.

Asia

Feeding Grounds

North America

Mating Grounds

Whales on Display

Breaching

Many whales perform a variety of unique movements. Three are shown here. After leaping halfway out of the water, a whale will twist in the air and splash noisily back down. This movement is known as breaching and may be done over and over again. Some scientists think it is a form of communication; others believe the animal is trying to loosen the whale lice and barnacles stuck to its body.

Lobtailing

Whales sometimes slap the water with their tail in order to communicate with one another, a behavior called lobtailing. In an aggressive version, a rival male whale will lift his tail high and then smack the water—or another male whale. This looks playful, but the force of this movement can injure the opponent.

Spy-Hopping

Sometimes a whale will stand on its tail and poke its head vertically out of the water. This fancy move is called spy-hopping. Again, scientists are not sure exactly why whales do this, although some say this gives the animal a good opportunity to "spy" on what's going on above water.

Blubber Blanket

How do whales keep warm in the icy-cold waters in which they live? They are, after all, **warm-blooded** mammals. Fortunately, nature has given them a very thick layer of blubber, or fat, below the skin. It can be as thick as 50 cm (20 in.). This layer insulates the whale and is a source of energy when there isn't much food around.

Blubber

Muscle

First Whales

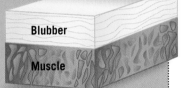

Scientists have found convincing evidence that the ancestors of whales were land mammals. A whale fossil was found in a part of Egypt that was once under the sea. The 40-million-year-old fossil is a whale leg, complete with kneecap, ankle, and toes! It is thought that over millions of years, whales lost their legs from lack of use.

Whales Close up

Whaling

Until a few hundred years ago, the world whale population was in the millions. Then, in the 17th and 18th centuries, whaling became a huge industry. Whales were killed mainly for their blubber, which was used for lamp oil and candles. Whale teeth and bones were carved into decorative items, like this scrimshaw jagging wheel. Unfortunately, some **species** were hunted until they were almost **extinct**. In 1986, commercial whaling was banned.

White Whales

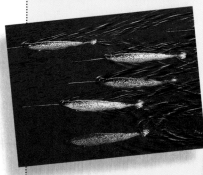

a long tooth that grows out through a hole in his upper lip; it can reach up to 3 m (10 ft.) in length. Some people believe these "horns" may have fueled stories about the mythical horned horse, the unicorn.

There are two species of white whale, the beluga *(right)* and the narwhal *(above)*.

Beluga whales were known in the 19th century as sea canaries for their variety of song-like sounds, which include squeals, whistles, and clicks.

The male narwhal has what appears to be a long horn growing out of his forehead. It is actually

Moby Dick

One of the greatest American novels is *Moby Dick* by Herman Melville, published in 1851. It tells the story of Captain Ahab's obsession with hunting down the great sperm whale, Moby Dick, who bit off his leg. The novel is full of symbolism and tells of the struggle between good and evil.

Melville himself worked on whaling ships. So when you read the novel you get a good idea of what it was like to be on board a whaling ship in the 19th century.

In the 1930 edition of *Moby Dick,* artist Rockwell Kent shows the great whale attacking a ship.

A Flower for Protection

When endangered, a group of sperm whales will assume a "lotus flower" formation. The herd gathers closely, putting their heads together in the middle and pointing their powerful tails outward to repel attackers. Young calves and injured adults are kept protected in the center of the formation. Few **predators** would brave the mighty thrashing tails and flukes.

Whale Song

During mating season, the male humpback whale makes long patterns of sounds that are loud enough to be heard for kilometers. Since the patterns are repeated, they can be called songs.

Dr. Roger Payne *(right)*, a whale biologist, has spent 30 years studying their songs and has developed a way of writing them down like musical notes. He finds that each song usually lasts about 15 minutes, but sometimes they go on for 24 hours or longer. In this way, whales communicate with one another over long distances.

The sounds they make are low, rather mysterious moaning and bellowing. Dr. Payne says these songs

"give the ocean its voice."

Whale songs were among the sounds of earth recorded on a videodisk that went into space on board the two *Voyager* space probes launched in the 1970s.

Killer Whales

The greatest hunters of the ocean are killer whales, or orcas. These striking black-and-white creatures hunt down fish, penguins, **pinnipeds**, sharks, dolphins, and porpoises. By hunting in a ferocious pack of up to 30, they even dare to attack larger whales like the blue whale.

Orcas have sharp rows of teeth that point backward, making it possible for them to tear chunks of flesh from their **prey.** Fortunately, they do not attack humans.

Whale Mask

The legends of the Kwakiutl, a Native American tribe of the Northwest, tell of supernatural beings called Ocean People who live under the sea. The most powerful of these beings is the killer whale. During rituals and celebrations the Kwakiutl pay tribute to this mighty ocean being by wearing this imaginative killer-whale mask *(left)*.

Tidal Zones

For plants and **animals** living along the seashore, life is anything but a day at the beach. High **tides** swamp them with water and bring in **predators** from the sea. Low tides expose them to the wind, sun, hungry birds, and souvenir-seeking beachcombers.

Many coastal creatures spend at least part of the day out of water, where they might dry out and die. To avoid this danger, some animals, such as barnacles and mussels, live inside protective shells. Sea stars and crabs retreat to tide pools or under rocks and seaweed. Sea anemones have a thick exterior that helps them stay moist.

A challenge faced by all plants and animals living between the tides, though, is staying put in the face of crashing waves and tugging tides—a problem that leads to some pretty sticky situations *(next page)*.

Life inside a tide pool is always changing. With every high tide, new plants and animals come in while others get out. That's why it's such an adventure walking along the shore at low tide peering into tide pools.

Mussels

Sea Star

(next page)

Would **You** Believe?

Tidal Freeloader

The sea anemone is a predator that stays in one place. It uses an adhesive foot to anchor itself to a hard surface. From there it waves its stinging tentacles to capture food that swims or floats by. Occasionally, though, anemones get creative, like the one below grabbing a piggyback ride from a hermit crab. By going mobile, the anemone increases its feeding range. The crab gets a bodyguard and, if it's lucky, even a few leftovers.

The Highs and Lows

Splash Zone
Rarely covered by water, this zone is wetted only by splashing waves and very high tides.

High Intertidal
A tidal area that is covered by nearly all high tides.

Middle Intertidal
This level is uncovered during every low tide.

Low Intertidal
A region that lies near or below the low-tide mark.

Red Algae

Sea Urchin

Sea Anemone

Sea Star

Imagine you are a fish living in a shallow tide pool. Sea gulls are circling overhead just looking for a little fish like you to make into a tasty meal. Luckily, fish that live in tide pools have eyes near the top of their head like frogs. This allows them to see the danger above and dart out of harm's way. The red-spotted blenny *(right)* is a good example.

Though small in size, tide pools contain an amazing variety of life. Plants flourish because there's lots of sunlight. They provide food for mussels and snails like periwinkles, which are eaten by sea stars, fish, and octopuses. Crabs scavenge for whatever's floating around.

Get a Grip

Tidal plants and animals must be able to hold on tight when hit by waves. Coastal kelp *(right)* uses a kind of root called a holdfast to grab the shore.

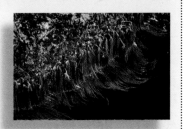

The chiton, a relative of the snail, has a muscular foot the size of its entire body that uses suction to fasten itself securely to rock.

Sea stars have hundreds of fluid-filled tube feet *(pages 50-51)*. The tips of the feet act like little suckers to grip rock and other hard surfaces.

Barnacles cement themselves headfirst to rock and then extend their legs to grab food floating by. At low tide they close up tight inside their shell to keep from drying out.

Coral Reefs

They may look like plants or rocks, but coral reefs are actually **colonies** of **animals.** Coral polyps, as they are called, are no bigger than your thumb and belong to the same **phylum** as jellyfish and sea anemones *(pages 38-39).* Yet these tiny creatures construct reefs that are larger than anything built by any other living animal, including humans. These reefs provide a perfect **habitat** for a very diverse group of animals *(pages 98-99).*

Even though corals aren't plants, they depend on plants for their survival. Each polyp has microscopic plants called **algae** *(pages 24-25)* living in its body. The algae benefit from this arrangement because they get **nutrients** and carbon dioxide to breathe from the coral. The coral in turn gets a built-in food source because algae, like all plants, make their own food through **photosynthesis.**

How Coral Is Created

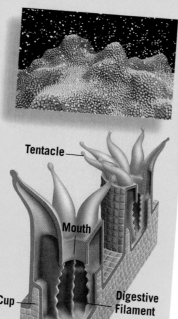

Like sea anemones *(pages 38-39),* coral polyps have a mouth surrounded by stinging tentacles. Each soft-bodied polyp makes a hard limestone cup, or **skeleton,** for protection. Individual skeletons fuse together to make reefs *(bottom).*

Coral reproduces in two ways. Baby polyps can bud off the parent polyp, or coral can reproduce sexually. Sexual reproduction, called spawning, can be a spectacular event. Once a year, some coral colonies spew millions of eggs and sperm into the water *(top).* **Fertilized** eggs then drift to new locations to settle and grow into adult polyps.

Tentacle

Mouth

Cup

Digestive Filament

Reef

Aquatic Garden

A healthy coral reef like Australia's Great Barrier Reef *(below)* is home to many different kinds of coral, both hard and soft. There are about 1,000 different species of hard coral and nearly 2,000 species of soft coral.

Soft and Hard Coral

Soft corals like sea fans and sea whips do not build reefs. They have flexible internal skeletons and live in colonies that project out into the water currents. There the polyps *(above)* are in the best position to grab passing food.

Hard corals build reefs in shallow coastal waters where there is lots of sunlight. There are three types of reef. Fringing reefs grow close to shore. Barrier reefs enclose lagoons between themselves and the shore. Coral atolls grow around underwater volcanoes.

Types of Reefs

Fringing Reef

Barrier Reef

Coral Atoll

Coral Reefs Close up

Coral reefs are more than just wonders of underwater architecture. They are an important source of tourist money for many countries. They support fish and other **animals** that feed millions of people. And they protect shorelines from storm surges and soil erosion. But by exploiting these advantages, humans have endangered many of the world's reefs.

Too many boaters and divers touching the reefs is just one reason why almost all the reefs in the Florida Keys *(right)* are considered endangered by the World Resources Institute. Other threats to reef survival include too much fishing, pollution, and building construction along the coast. Nature can be destructive, too. Some sea creatures eat the reefs. Certain weather patterns *(pages 22-23)* can increase water temperature to the point that **algae** living inside coral polyps flee, leaving corals without their main food source. Between natural and man-made stress, more than half of all coral reefs are now considered at risk.

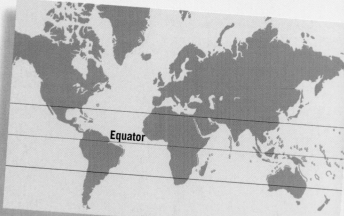

Equator

Where Coral Reefs Are Found

All coral reefs *(shown in red)* can be found in a narrow band 30° above and below the equator. Here the waters are shallow and clear, exposing coral to lots of sunlight—a necessity for the algae living inside coral that need sunlight for **photosynthesis.**

Most types of coral live in the Indian and Pacific Oceans around Southeast Asia. The rocks around the many islands and underwater volcanoes there provide a good surface for coral to attach to and grow.

Coral Reefs in Danger

A diver in the Philippines earns a living by harvesting sea urchins and chunks of coral. Overfishing, though, is one of several threats facing reefs. High demand for coral reef fish as food and as aquarium pets has led to two destructive tactics: blast fishing and cyanide fishing, which use underwater explosives and poison to stun fish for easy capture.

How **Big?**

The Great Barrier Reef

If corals are the world's master builders, the Great Barrier Reef is their masterpiece. Stretching for 2,000 km (1,260 mi.) along Australia's northeast coastline, the reef is almost as long as one of mankind's greatest building projects: the Great Wall of China. But whereas the Great Wall is 2,400 km (1,500 mi.) long, it is only 7.5 m (25 ft.) high, it is only 7.5 m (25 ft.) high. The Great Barrier Reef is an astounding 150 m (500 ft.) high, making it the largest single structure ever built by living creatures.

Great Barrier Reef—

Chomp!

The parrotfish uses a pair of powerful beaklike jaws to bite off chunks of algae-rich coral. It also uses its sharp teeth to scrape algae from coral surfaces.

Crunch!

The crown of thorns sea star is a thorn in a coral reef's side. These massive, prickly echinoderms *(pages 50-51)*— sometimes 45 cm (18 in.) wide —can devour large sections of reefs in a short period of time. In the Pacific and Indian Oceans, crown of thorns populations have devastated a number of reefs.

Yuck!

It may not be pretty to you, but the hairy striated frogfish looks like a delicious algae-covered rock to unsuspecting small fish. Because of its not-so-good looks, the frog-fish enjoys the advantage of letting its **prey** come to it, rather than having to hunt.

Coral Citizens

Coral reefs are like undersea cities. The millions of nooks and crevices along sprawling reefs offer **marine** animals of all shapes and sizes places to hide, sleep, feed, and breed. In fact, at least a fourth of all known marine **species** live in reefs. Australia's Great Barrier Reef *(page 97)* alone is home to 2,000 different kinds of fish.

As in any city, the citizens of a reef operate on different schedules. Some are early risers, others are night owls. During the day the city explodes with color. The bright hues and patterns of tropical fish may serve to camouflage them as they feed in sunlit waters. Night feeders tend to be duller, the better to lurk in dark alleys for **prey.**

Anemone Fish
The sting of a sea anemone doesn't bother this tiny fish, which nestles in the deadly tentacles to protect itself. In return, the anemone fish helps the anemone by eating **parasites** and driving away **predators.**

Cleaner Wrasse
Teeth or gills need cleaning? Stop by a cleaning station set up by a type of fish called a cleaner wrasse. Fish line up and wait their turn to have fragments of food and harmful parasites picked away by cleaner wrasses.

Nudibranch
The bright neon colors of this sea slug, called a nudibranch, visually scream a warning to predators: "Don't eat me, I taste really bad!"

Sea Horse
A 6.25-cm (2.5-in.)-long pygmy sea horse camouflages itself in the branches of a variety of soft coral. Each kind of pygmy sea horse is perfectly matched to one type of coral.

Night Shift

Octopus

The octopus has extremely good eyesight *(page 45)*, a real advantage when it is crawling around a reef at night hunting for food.

Squid

The soft-bodied squid uses intelligence and speed as well as the cover of night to hunt for food safely along the reef.

Moray Eel

A cleaner shrimp has no fear as it picks the sharp teeth of a scary-looking moray eel. At dusk, these snakelike creatures emerge from crevices to search for food.

Sea Worm

Although only a few centimeters long, the segmented sea worm captures fish with serrated jaws that are twice as wide as its body.

Let's Compare

Daylife vs. Nightlife

During the day, a coral reef looks like lifeless rocks. The thousands of tiny coral animals that make up the reef are closed up tight *(top)*. At night, though, the reef wriggles to life as coral polyps extend their tentacles to feed *(bottom)*. Like their close relatives, sea anemones, coral polyps are stationary feeders, snaring particles of food as they drift by.

The Deep Sea

Fabulous Features!

As you go deeper into the ocean, the amount of light decreases and the strangeness of the **animals** increases.

At a depth of 180 m (600 ft.) you enter the twilight zone. Many creatures in this dusky realm provide their own light—a process called **bioluminescence.** Some produce a glowing substance called luciferin. Others borrow light made by other creatures, either by eating luminous organisms or by playing host to light-producing **bacteria.**

Below 1,000 m (3,300 ft.) you enter the dark zone. Because no sunlight reaches this deep, there are no plants. This makes for very little food. So creatures living here generally have a huge mouth and large, sharp teeth so they can more easily grab whatever floats by. They also tend to have a soft, flexible body that won't be crushed by the enormous weight of the water above.

Animals that live in the twilight zone and below must contend with three major challenges: extreme water **pressure**, the absence of light, and the scarcity of food. As a result, the deep sea is populated with all sorts of large-mouthed, glow-in-the-dark creatures.

Soft Body
The deep-sea swallower has enormous jaws for capturing **prey** and a soft, compressible body that cannot be crushed by enormous water pressure.

Bioluminescence
Flashlight fish have **organs** under their eyes filled with glowing **bacteria,** allowing them to see in the dark.

Big Teeth
The snakelike fangs of the viperfish are almost too big for its mouth—perfect for spearing scarce food whenever it floats by.

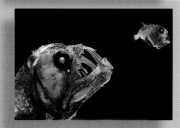

How Loud?

Scientists once thought that fish made no noise at all. Were they ever wrong about that! What look like a gang of goblins below are **embryos** of the midshipman fish. They came into being because of a strange—and very loud—mating ritual. In late spring, midshipman males build nests on the seafloor near the coast of North America. To attract females, they hum. The humming, said to sound like the chanting of monks, is so loud that it can be heard on the surface.

The ghoulish anglerfish uses trickery to lure fish into its dagger-filled jaws. A kind of spine poking up from the front of its head acts like a fishing pole. Glowing tissue at the end of this long stalk attracts fish like bait.

What Big Eyes You Have!

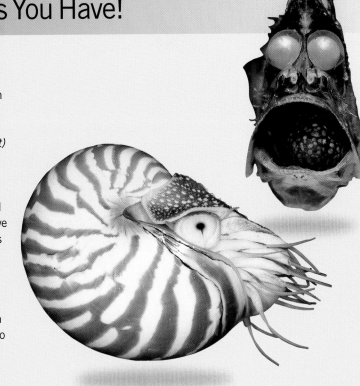

It helps to have big eyes when you are trying to get around in the dark waters of the deep ocean. The hatchetfish *(far right)* uses a pair of bulging lenses to detect bioluminescent fish, which it then captures in its oversize mouth. The chambered nautilus *(near right)* is a reclusive relative of the squid. Unlike its cousin, which has no external protection, the nautilus lives in the outermost chamber of a 36-chamber coiled shell, approximately 25 cm (10 in.) in diameter. It uses its large eyes to locate food in the seas around the East Indies and Fiji Islands.

How Deep?

How deep in the ocean can you go before light disappears? As far down as 180 m (600 ft.) there is enough sunlight for plants to grow. This is called the sunlit zone. From there you enter the twilight zone. Only blue light penetrates this deep. No light at all reaches below 1,000 m (3,300 ft.). So the part of the ocean that gets any light is a very tiny layer indeed!

Sunlit Zone
0-180 m
(0-600 ft.)

Twilight Zone
180-1,000 m
(600-3,300 ft.)

Dark Zone
1,000-4,060 m
(3,300-13,200 ft.)

Abyss
4,060-6,092 m
(13,200-19,800 ft.)

Empire State Building, New York 381 m
(1,250 ft.)

Eiffel Tower, Paris 300 m
(984 ft.)

Then & NOW!

Living Fossil

A diver *(above, top)* swims alongside a relic from prehistoric times —the coelacanth. Coelacanths first appeared on earth about 400 million years ago, nearly 200 million years before dinosaurs. The large bottom-dwelling fish was believed to be extinct until 1938, when one was caught in the Indian Ocean near Madagascar. Since then, scientists have assumed that what few coelacanths remained lived in this one location. But the one shown above was found thousands of miles away in waters around the islands of Indonesia. Fossils *(above, bottom)* reveal that coelacanths have remained relatively unchanged over the past 400 million years.

Deep-Sea Floor

Oasis of the Deep

The deep-sea floor was once thought to be a desert, without any life. What creature could survive in the dark, freezing waters under tons of **pressure?**

When vessels were finally invented that allowed us to explore this realm—known as the abyss—we discovered a strange new universe. **Animals,** forced to live on food particles that filtered down from 6,000 m (20,000 ft.) above, had become either passive drifters or nearly motionless to conserve energy. Instead of bones that could be crushed by the enormous pressures, their bodies were found to contain soft **cartilage,** fat, or liquid. In total darkness, some had no use for eyes or body **pigment.**

In places, volcanic activity within the earth causes mineral-rich hot water to erupt through fissures into the freezing water above. These hydrothermal vents are like oases of the deep sea, allowing chemical-based **ecosystems** to thrive in an otherwise uninhabitable environment.

What's Chemosynthesis?

Where there is sunlight there are plants that turn the sun's energy into food through **photosynthesis.** These plants then become food for animals. But what happens at the bottom of the ocean, where there is no sunlight? In this lightless realm **bacteria** use chemical energy to make food. This process is called chemosynthesis. Chemosynthesis makes life possible near the ocean floor. Animals graze on bacteria, which form the foundation of the deep-ocean food chain *(right).*

Fish and Crabs

Limpets

Tubeworms

Bacteria

Cracks in the seafloor, called hydrothermal vents, spew hot water into the freezing ocean depths through large chimneys. These chimneys, called black smokers, emit mineral-rich water that feeds bacteria. These bacteria form the basis of a diverse ecosystem on the ocean floor. Clustered around most vents are giant tubeworms that grow up to 1.8 m (6 ft.) long. Tubeworms are stationary and have no mouth or gut—bacteria live within the tubes, providing food in exchange for a place to live.

Alvin

The U.S. Navy submersible *Alvin* helped revolutionize deep-sea exploration. Able to carry a crew of three to depths of 4,000 m (13,000 ft.), *Alvin* can explore about 40 percent of the seafloor with cameras and a robotic arm. *Alvin* was the first to discover life thriving kilometers below the surface when its crew explored a mid-ocean ridge near the Galápagos Islands in 1977. They revealed an unusual **habitat** sustained by thermal vents *(illustration below)*.

Ocean Mining

Would You Believe?

The ocean floor has long been mined for oil and manganese, two of its most valuable deposits. Recent discoveries of gold, though, have provoked a modern-day gold rush to the bottom of the sea. In 1994, British scientists drilled into a volcanic mound 3.2 km (2 mi.) deep in the Atlantic Ocean. It contained five times as much gold as most deposits mined on land.

This 20-cm (8-in.)-long deep-sea isopod is related to **crustaceans** *(pages 48-49)* but looks more like a giant cockroach scurrying along the ocean floor.

Two sea anemones anchored to a glass sponge resemble an exotic plant. Glass sponges are survivors from the time of dinosaurs.

The eyeless deepwater sea spider suspends itself just above the ocean floor—5,000 m (16,500 ft.) down—to feed from passing currents.

Polar Extremes

Antarctica and the Arctic Circle are two of the harshest **habitats** on the planet. Icy waters, freezing temperatures, and alternating seasons of total light and total darkness are forbidding to most plants and **animals.** Despite severe conditions, people and wildlife live year round in the Arctic, which is actually a frozen ocean. Antarctica, though, is like another planet. Earth's fifth largest continent is home to the world's largest desert, the most active lava lake, the strongest winds—320 km/h (200 mph)—and the coldest recorded temperature— –89.5°C (–128.6°F). No one lives there permanently, and the water is too cold for most fish. Yet the waters are surprisingly rich with life, in large part because an annual upwelling brings **nutrient-**rich waters up from the bottom of the ocean.

North Pole

South Pole

Penguins

An emperor penguin goes airborne after launching itself through the ice after a swim. Though they must go to the sea to feed on krill, squid, and fish, emperors waste no time returning to land because of **predators** like the leopard seal lurking in the water. There are 17 kinds of penguins, and most live on islands and in the icy water near Antarctica. The aptly named emperor is the largest. These penguin giants can grow to be 115 cm (45 in.) tall. They are the only **warm-blooded** animal that lives year round on Antarctica.

Seals

Crabeater seals are one of five species of pinnipeds that live in the icy waters around Antarctica. Despite their name, these seals feed mainly on tiny shrimplike krill, the primary food for many animals near the South Pole.

Arctic Terror

This is the view enjoyed by seals when a polar bear goes fishing. The massive front paws are dangerous weapons; they are also partially webbed, one reason polar bears are good swimmers. A thick layer of fat and dense fur keep these bears warm whether they're on land or in the water, and they are experts at conserving energy. Polar bears can fast for months at a time when food is scarce, and their remarkable sense of smell allows them to hunt seals by lying in wait next to a seal's breathing hole.

Strange But TRUE!

Blood That Won't Freeze

The ghostly icefish is one of just a few **species** of fish that can live in the freezing waters around Antarctica. Its pale appearance is due to a lack of hemoglobin, a reddish brown protein found in the blood of other **vertebrates**. Instead the icefish is blessed with a special compound in its body fluids that acts like a natural antifreeze. Bulging eyes and a huge mouth are other bizarre-looking but beneficial features in this stark environment.

I Was There!

Becoming the first person to photograph the elusive Greenland shark in the wild was sometimes a frightening experience for Nick Caloyianis (*above*), especially when he witnessed the shark's ability to capture food from a distance. "We lured one with a three-foot mesh bag of bait," says Caloyianis. "The shark inhaled it from three feet away. I quickly aimed my camera, focused, and clicked off two shots—all that my half-frozen equipment could manage." Despite being one of the world's largest sharks, and the only one in the Arctic, the Greenland shark has rarely been studied because of how deep it lives in the icy Arctic waters.

Ancient Mariners

Odysseus

The oceans were a mysterious and dangerous place for sailors of long ago, who had no maps to guide them. Their ships were powered only by sails. At any moment, a great storm might swoop down and send them to a watery grave or ship-wreck them on a distant island. Many ships and their crews were away from home for years. Or they never returned at all.

These early seafarers feared imaginary dangers in the unknown oceans. The gods and goddesses who ruled the waves might, in a moment of anger, whip the ocean waters into a deadly whirlpool or typhoon. Sailors believed in ferocious sea monsters, as well. Some had gruesome heads with poisonous fangs, or razor-sharp claws that could rip a wooden ship to shreds. Tales of these beasts probably grew out of sightings of real sea creatures, such as the oarfish *(next page)* and giant squid.

The most famous sea traveler of ancient Greece was the mythical king Odysseus. His story is told in a long poem written by the Greek poet Homer called the *Odyssey*. The king was forced to wander the seas for 20 years because he had angered Poseidon. He faced many dangers, including the cliff monster Scylla *(above)*, whose six heads on long necks could reach down into the water below. Scylla fed on dolphins and sharks, and sometimes on sailors from passing ships.

Poseidon

The ancient Greeks believed the oceans were ruled by a terrifying god called Poseidon. (The Romans called him Neptune.) He lived in a cave deep under the sea just east of Greece. When he was angry, he unleashed floods and sea storms upon helpless humans.

The Greeks believed Poseidon traveled across the water in a chariot drawn by 100 white horses with golden manes and brass hoofs. As he raced over the sea, Poseidon always held a trident, or three-pronged spear.

The Mythical Voyage of Saint Brendan

Saint Brendan was a monk who lived in Ireland about 1,500 years ago. According to Irish folk tales, Saint Brendan and 17 other monks set out in a leather boat to find the legendary Land Promised to the Saints. They came across all kinds of magical islands and sea monsters. They tamed one whale long enough to hold Easter services on its back *(left)*. The legend says that Saint Brendan eventually found the promised land and brought home jewels from its golden shores.

Imagine That!

Sea Monsters

For centuries, sailors have told tales of strange and terrifying dragons *(below)*, sea devils *(above, right),* or huge whales that disguised themselves as islands *(below, right).* Scandinavian sailors claimed sightings of an enormous horned monster—the Kraken *(right)*—that had many heads and tentacles and made an inky, poisonous liquid. It was probably an exaggerated version of a real-life creature—a giant squid.

Real Monsters

Many of the sea serpents sighted by sailors were probably oarfish *(right)*. These silvery, ribbon-shaped fish can be more than 6 m (20 ft.) long. One person claimed to have seen an oarfish that was more than 15 m (50 ft.) long! The oarfish swims through the water like a snake. It lives very deep in the ocean and is rarely seen except when it is either dead or dying and comes to the surface. The oarfish has a bright red fin along its back that resembles a mane, which may explain why the sailors thought it was a fiery sea serpent.

Stick Maps

The first sailors steered by the stars and ocean currents and by watching migrating birds and fish. The Polynesians of the South Pacific were one of the earliest peoples to create sea charts. They made their maps out of thin bamboo sticks *(below)*. Each stick stood for an ocean current or sailing route. Shells or stones showed the location of islands. Skillful Polynesian sailors used these stick maps for thousands of years.

Challenger First Deep-Ocean Explorer

On December 21, 1872, the British research ship HMS *Challenger (right)* set sail on the first major expedition to study the ocean depths. The scientists on board were looking for answers to many questions: How deep are the oceans? Is there life in the deepest parts? Are all the world's seas equally salty?

The *Challenger*'s voyage lasted three and a half years. The ship zigzagged across three oceans, traveling more than 100,000 km (60,000 mi.). It collected 1,440 samples of water and 13,000 different kinds of **animals** and plants, including thousands that had never been seen before. It also dredged hundreds of samples of mud and rock from the seafloor. The *Challenger*'s scientists spent the next 20 years studying the information they had gathered. When finally published, it filled 50 very thick books!

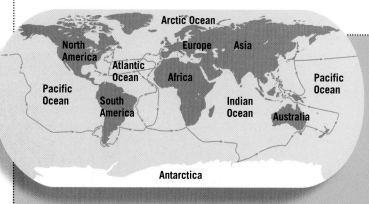

Arctic Ocean
North America
Europe
Asia
Atlantic Ocean
Africa
Pacific Ocean
Pacific Ocean
South America
Indian Ocean
Australia
Antarctica

Challenger's Route

As you can see from the map above, the *Challenger* explored every ocean in the world except the Arctic. During its long voyage, the ship encountered several fierce storms, and the scientists and crew had a few nerve-racking experiences. While sailing the frigid waters near Antarctica, the 2,100-t (2,300-tn.) *Challenger* crashed into an iceberg! Fortunately, the damage could be fixed, and the ship continued its historic voyage.

Collecting Samples

The *Challenger*'s crew used a special scoop called a dredge net to gather specimens from the ocean floor. The net was at the end of a very long rope that had a weight attached partway down *(near right)*. The weight pulled the dredge to the bottom. The movement of the ship filled the net with mud and deep-sea creatures.

Hundreds of samples, including sea anemones, jelly-fish, sea slugs, and worms, were sorted into different containers for later study *(far right)*. These specimens proved for the first time that life existed deep under the sea. One of the most surprising discoveries was the spirula, a squidlike creature that was thought to have died out 50 million years ago.

The *Challenger* had a crew of 20 officers and about 200 seamen. The ship captain was George Nares *(seated, third from right)*. The six scientists on board were led by biologist Charles Wyville Thomson *(seated, fifth from right)*. A professional artist, J. J. Wild *(standing behind Thomson)*, came along to make drawings of the voyage.

Famous 1 FIRSTS

Floating Laboratories

The *Challenger* was originally a warship. All but two of its 17 guns were removed to make room for scientific laboratories. The tiny chemistry laboratory had several burners, a sink, glassblowing equipment, and storage space for hundreds of chemicals. The naturalists' laboratory *(left)* included microscopes to examine fossils, shells, and fish specimens, a hydrometer to measure the relative **density** of water, and special thermometers that could record temperatures at the ocean's bottom. Thousands of drawings were made from the specimens collected from the ocean, including those of fish *(left, bottom)* and mollusks *(left, top)*.

Early Divers

Diving Suits

For centuries, divers in Japan and Korea have practiced holding their breath long enough to gather pearls and sponges off the bottom. But the record for this kind of dive is less than four minutes, not long enough to do any exploration.

As early as 333 BC, people began building special equipment to stay underwater for longer periods. That was the year, according to legend, that Alexander the Great was lowered into the ocean inside a waterproof glass barrel. By the 16th century, treasure hunters had invented diving bells to recover goods from sunken ships. Later, special suits designed for the same purpose let divers move about freely. The first workable submarines allowed people to travel underwater without ropes, air-hoses, or heavy helmets.

1700s

The first diving suits were heavy metal hoods that fitted over the head and upper body. They had a small glass window for the diver to look through. Air reached the diver through a long, flexible hose. These awkward suits could be used only for short dives at shallow depths.

1840s

Many diving suits of the mid-1800s made the wearer look like an alien from outer space! The copper-and-glass helmets weighed up to 9 kg (20 lb.). The suits were made of thick waterproof canvas. Air was pumped down from the surface through long pipes attached to the helmet.

Diving Bell

In 1690, the British astronomer Edmund Halley built one of the first successful diving bells. It was a wooden drum coated with lead and had a glass top. Fresh air was lowered to the bell in barrels. By the mid-1800s, diving bells like the *American Nautilus (above)* used air pumped in under pressure from a boat on the surface.

1900s

Divers who used this suit in the early 1900s still needed to have air pumped down to them through long tubes. But the suit's helmet had an automatic valve that controlled the amount of fresh air that came in. Just as with earlier diving suits, the boots were weighted with lead to make it easier for the diver to walk underwater.

Early Submarines

One of the most amazing early submarines was the *Turtle (left),* built in 1776. The single crewman made the craft sink or rise by pumping water from its tanks. During the American Revolution (1775-1783), the *Turtle* tried to attach explosives to a British warship, but the attack failed.

Around 1800, American inventor Robert Fulton built the *Nautilus,* a small iron submarine powered by a hand crank. A German named Wilhelm Bauer designed a submarine in 1855 that used men on treadmills to turn the propeller. He sold it to Russia where, it is said, a small orchestra played aboard it during the 1856 coronation of Tsar Alexander II *(below)*—the first underwater concert!

Bathysphere

Underwater explorer Charles William Beebe emerges from the bathysphere that he and another American, Otis Barton, invented in 1930. A hollow steel ball with thick quartz windows and powerful lights, the bathysphere (*bathy* is Greek for "deep") was lowered from a ship by heavy cables. It could move only up and down. In 1934, Beebe and Barton took the bathysphere down to 923 m (3,028 ft.)— the deepest that anyone had gone at the time.

Imagine That!

Attack of the Giant Squid

Early submarines inspired French writer Jules Verne's famous 1870 novel *Twenty Thousand Leagues under the Sea.* The book describes the exciting adventures of Captain Nemo and his crew aboard the fabulous submarine *Nautilus.* It could travel at the unheard-of speed of 80 km/h (50 mph). One of the best-known scenes involves a battle with a giant squid *(right).* Verne's imaginary creature weighed a whopping 23,000 kg (50,000 lb.)!

Exploring the Depths

Jacques Cousteau

Exploring the deepest parts of the ocean requires special equipment—much like the astronauts use in space. One of the biggest dangers facing deep-sea divers is the enormous **pressure** of the water above. A diver's lungs can easily be crushed by pressure if protective equipment isn't worn. And if a diver returns to the surface from a deep dive too quickly, bubbles of nitrogen can form in the body. This causes the bends, a painful condition that can kill the diver *(page 13)*.

Scuba equipment allows a diver to safely explore to depths of up to 90 m (300 ft.). The portable air tanks have a special valve that supplies the diver with just the right amount of air needed for each breath. Many scuba divers carry small computers to tell them how long to remain at certain depths to avoid the bends.

To go deeper, divers need to be in submersibles or pressure-proof diving suits. The newest submersibles can take people down to the incredible depth of 6 km (3.7 mi.). These vehicles allowed scientists and historians to explore the final resting place of the *Titanic* 4 km (2.5 mi.) beneath the surface of the Atlantic *(page 116)*.

Before his death in 1997, France's Jacques Cousteau *(below)* was the most famous oceanographer in the world. In 1943, he and a friend, Emile Gagnan, invented the self-contained underwater breathing apparatus, or scuba. With scuba gear, divers can swim freely underwater for several hours. Cousteau also built the *Diving Saucer (above)*, a two-person submersible that used jet nozzles to steer in any direction. It could dive to a depth of 300 m (1,000 ft.) and traveled at a speed of about 2 km/h (1 mph).

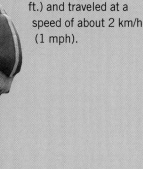

Fast FACTS

333 BC Alexander the Great is lowered into the ocean in a glass barrel, becoming the first undersea explorer.

1535 Italian Guglielmo de Lorena invents the first diving bell.

1620 Dutchman Cornelis Drebbel builds the first workable submarine.

1930 Charles William Beebe and Otis Barton invent the bathysphere.

1960 Using the submersible *Trieste,* Jacques Piccard and Don Walsh are the first to descend to the deepest spot on earth.

1979 Wearing a JIM suit, American oceanographer Sylvia Earle breaks the record for deepest underwater dive unconnected to the surface—380 m (1,250 ft.).

JIM Suit

Looking like an astronaut, a diver explores the ocean floor in a JIM suit. (The first person to test it was named Jim.) These high-tech suits are sometimes called personal submersibles and can reach a depth of 600 m (2,000 ft.). The air pressure inside the suit stays the same as at **sea level**, no matter how deep the diver goes. That means the person wearing the suit doesn't have to worry about getting the bends when surfacing from a dive.

Exosuit

In an Exosuit, a diver can swim freely—not attached to any hoses—at depths of up to 180 m (600 ft.)—deeper than with scuba gear. Made out of lightweight fiberglass, the Exosuit weighs only 73 kg (160 lb.). Its tanks carry enough compressed air to keep a diver underwater for two days. The Exosuit has a built-in communication system so the diver can talk to support crew on the surface.

Deep Worker

The one-person submersible *Deep Worker* is so small it can fit in the back of a van. Divers say steering the vehicle is as easy as driving a car—only a lot slower. Its top speed is only 7.4 km/h (4.6 mph)! *Deep Worker* can take its operator to depths of up to 600 m (2,000 ft.). But it is often used at shallower depths to help film underwater scenes for movies such as *The Abyss* and *Titanic*.

I Was There!

At about 1,000 feet (300 m), a region of deep-sea dunes begins. I head the sub on a course for deeper water. . . . At 1,500 feet (460 m), a huge rock looms in my path. . . . From the edge of the rock a large fish —a grouper!—appears, saunters directly over to [the sub's] dome and peers in. Light this bright must never before have shone into the life of this fish. . . . Certainly no submersible has previously come to call. . . . Two sets of eyes meet. "Who are you?" I want to know. "What is it like to spend your time where day and night subtly merge, where sea worms dance, where the sea is cold and you swim with the weight of forty-six atmospheres pressing down on your scaly hide?"

—Oceanographer Sylvia Earle, describing her 1986 dive in a one-person submersible

Sea-Link Submersible

Modern submersibles like the *Sea-Link (right)* make it easier to explore the ocean's depths. To enter the **pressure**-proof chamber, divers squeeze through a tiny hatch that is only 47 cm (18.5 in.) wide. The clear dome lets three passengers and the pilot see all around them.

The *Sea-Link* is lowered into the water from a support ship. It can go down to 1,000 m (3,000 ft.). Not much sunlight reaches that deep, so the *Sea-Link* has very powerful lights. But even with the lights, the divers can't see much farther than about 9 m (30 ft.). The pilot has to use a compass and other equipment to keep from getting lost.

The *Sea-Link*'s robotic arms and suction devices can capture creatures that live on or near the ocean's floor. In 1997, *Sea-Link* scientists discovered a rare deep-sea octopus whose suckers glow in the dark!

Try it!

How Much Is Left to Explore?

Although most of the ocean's floor has been mapped, very little of it has been explored. That's because the ocean covers more than two-thirds of the earth's surface. Here's a way to demonstrate just how much of the ocean is waiting to be explored: Fold a piece of notebook paper in thirds. Two-thirds of the paper represents the earth's oceans, and one-third represents the land. Now place a U.S. quarter or a bottle cap on the paper. That represents the part of the oceans already studied. Everything else is unexplored territory!

How Deep?

Trieste

The *Trieste* was the first submersible to reach the deepest spot on earth—the Mariana Trench in the Pacific Ocean. In 1960, Swiss engineer Jacques Piccard and U.S. Navy lieutenant Don Walsh took the *Trieste* 11,035 m (36,205 ft.) to the very bottom of the trench. They discovered that creatures and plants can live even in the deepest sea under incredible pressure.

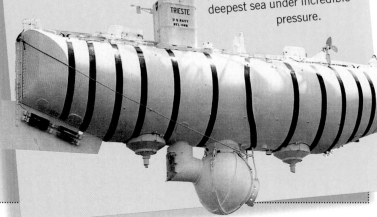

Heads Up!

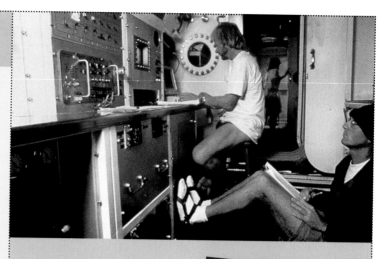

Waves and wind can make it difficult for oceanographers aboard ships to take measurements and conduct experiments. The Floating Instrument Platform (FLIP) helps solve that problem. After FLIP is towed to a research site, the crew floods its long hull with water. Within 20 minutes, the hull sinks stern first *(above, right),* leaving the bow section on the surface. The scientists and crew are able to live and work in the four-story FLIP laboratory *(below)* for weeks at a time. Even if hit by a 9-m (30-ft.) swell, FLIP will move up or down no more than 1 m (3 ft.).

Living and Working Underwater

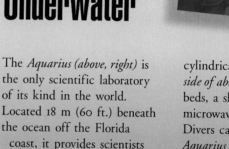

The *Aquarius (above, right)* is the only scientific laboratory of its kind in the world. Located 18 m (60 ft.) beneath the ocean off the Florida coast, it provides scientists with advanced research equipment and living accommodations. Its cylindrical chamber *(top left side of above photo)* contains beds, a shower, a toilet, a microwave, and a refrigerator. Divers can exit and reenter *Aquarius* to explore the nearby ocean floor without having to surface and risk getting the bends *(page 13).*

Underwater Hotel

Would you like to sleep beneath the sea? Just book a room at Jules' Undersea Lodge, the world's first underwater hotel. To reach the lodge, which was named after French novelist Jules Verne, you have to use scuba gear. You dive 9 m (30 ft.) down in a small, protected bay off the Atlantic coast of Florida. Inside are all the comforts of home, including a hot shower, a kitchen, a television for watching videos— and big windows with fascinating underwater views *(below).*

Sunken Ships and Treasures

The ocean floors are littered with the remains of wrecked ships and their cargo. Raging storms, jagged rocks, fog, mist, and ice have claimed countless ships ever since humans first started building boats. Even today with satellite communications, radar, and modern steel hulls, ships still can—and sometimes do—sink.

Some of the world's oldest shipwrecks, found in the Mediterranean, carried cargoes of copper bars, bronze tools, and pottery jars. Many other treasures, including gold, silver, jewelry, and works of art, have been lost in the centuries since.

Until recently, shipwrecks that lie on the ocean floor were out of reach of both oceanographers and treasure hunters. Today, special suits, submersibles, and underwater robots have made it possible for people to explore and salvage where it was never possible before.

"Then, directly in front of us, there it was: an endless slab of rusted steel rising out of the bottom—the massive hull of the *Titanic!* I felt like a space voyager peering at an alien city wall on some empty planet."

—Robert Ballard

Graveyard of the Atlantic

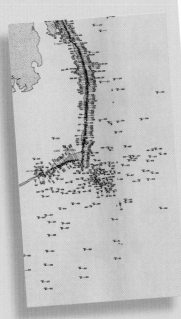

The rough seas and shifting sand bars along North Carolina's Outer Banks have earned that area the frightening nickname graveyard of the Atlantic. The map at right shows the location of 373 of the almost 700 ships that have wrecked there over the last four centuries. Some can still be seen at low **tide** along the shores of these low barrier islands that separate the mainland from the Atlantic Ocean.

Titanic

In April 1912, the British luxury liner *Titanic* struck an iceberg on its first, or maiden, voyage. The ship sank in the 4-km (2.5-mi.)-deep Atlantic, and more than 1,500 of the 2,227 passengers died. In 1985, ocean explorer Robert Ballard *(above, right)* and a team of American and French oceanographers located the wreck. The next year he descended to the site in a submersible named *Alvin* *(illustration above)*. Attached to *Alvin* was a remote-controlled robot submersible called *Jason Jr.* From inside *Alvin* the crew could send *Jason Jr.* into places too small or too dangerous for the larger sub.

Atocha

In 1622, a violent hurricane broke apart the Spanish treasure ship *Nuestra Señora de Atocha* off the coast of Florida. Gold bars, silver coins, and jewels were scattered across the ocean floor. For more than three centuries the *Atocha* lay undisturbed. Treasure hunter Mel Fisher *(far left)* spent 17 years searching for the wreck. Finally, in 1985 he found the main part of the ship. Using special salvage equipment *(top left)*, he and his crew recovered a treasure worth about $400 million. A model of the *Atocha* appears at bottom left.

Like many other sunken ships, the *Titanic* is being taken over by marine life. In the photo above, a feathery sea pen pokes out of a once elegant chandelier.

Ship of Gold

The California gold rush of 1848 set off a flurry of shipping activity from the West Coast to the East. The most direct route required two ships. The first brought the gold from California to Panama City, Panama, where it was moved across the narrow country by land. The second ship sailed from Aspinwall (now called Colón), Panama, to New York.

The U.S. mail steamship *Central America* had successfully completed 43 Panama-to-New York voyages. But on September 12, 1857, the ship sank during a hurricane off the Carolina coast. Three tons of gold were lost and 425 people died. In 1989, a team of engineers and scientists located and recovered the ship's valuable cargo.

The paddle-wheel steamer *Central America* sank at an angle of 45 degrees *(above)*. Passengers were thrown overboard into the churning water. The map at right shows the shipwreck site off the Carolina coast, where gold coins, bricks, and nuggets were scattered.

North America
Shipwreck Site
Aspinwall, Panama
San Francisco, California
Panama City, Panama
South America

Alexandria The Lost City

Two thousand years ago, Alexandria was the most important city in Egypt. Founded by Alexander the Great in 332 BC, Alexandria became a busy Mediterranean seaport and a center for learning. It had a huge library and a lighthouse that was considered one of the Seven Wonders of the World. Cleopatra, queen of Egypt, made her home in Alexandria.

Over the centuries, earthquakes and floods have caused the ancient buildings to sink beneath the waves. In 1961, an Egyptian diver discovered a collection of statues. After more ruins were discovered in 1993, salvage operations were launched. Thousands of artifacts were uncovered, including columns, giant statues, a dozen sphinxes, and pieces of the lighthouse. In 1998 it was confirmed that the palace of Cleopatra had been found.

The diver above records information about the wooden planks from an ancient ship. Each plank is carefully tagged and numbered.

Salvage divers pinpoint the exact location of artifacts by using a receiver linked to a network of satellites circling the earth *(above)*. The diver at left examines hieroglyphs, or word pictures, inscribed on a large granite block. The hieroglyphs describe a king "endowed with life and power." The diver at right goes head to head with a sphinx.

This gold ring was found near the site of an ancient shipwreck. The engraved stone in its center is made of chalcedony, a type of blue quartz.

This granite statue of the priest of Isis is more than 1.5 m (5 ft.) tall. The priest is holding a jar in the shape of a human head, which was considered a sacred image.

At right, two amphorae—or dual-handled clay jars—are tagged for identification purposes. Amphorae were used by early Mediterranean cultures to hold such food items as olive oil, olives, and wine.

Picture Credits

The sources for the illustrations in this book appear below. Credits from left to right are separated by semicolons, from top to bottom by dashes.

Cover: Book spine, © Michelle Hall/HHP; front, art by Jerry Lofaro (background); © Howard Hall/HHP; CORBIS/Rick Doyle—© 1999 Nuytco Research Ltd., North Vancouver, B.C.

3: Art by Will Nelson. **4:** © Ned DeLoach; © 1999 Norbert Wu/www.norbertwu.com—art by Will Nelson. **5:** Art by Stephen R. Wagner; Discovery Channel—© 1995 Sanford/Agliolo/Stock Market. **6, 7:** Charles Nicklin/Al Giddings Images, Inc.; © Roger Steene. **8, 9:** Map by John Drummond; © David Doubilet; CORBIS/Phil Schermeister—from *Atlas of the Oceans,* by John Pernetta (Mitchell Beazley, 1977), © George Philip Ltd., London; National Maritime Museum, Greenwich, London. **10, 11:** NASA; illustration by Oliver Rennert, © 1995 Weldon Owen Pty. Ltd. (background); © David Johnson; Dr. P. R. Bown, Department of Geological Sciences, University College London, London—World Ocean Floor, Bruce C. Heezen and Marie Tharp, 1977, © Marie Tharp 1977. Reproduced by permission of Marie Tharp, South Nyack, N.Y. (2). **12:** © Gerry Ellis/Minden Pictures—CORBIS/The Purcell Team. **13:** © Flip Nicklin/Minden Pictures; © David Doubilet—CORBIS/James Marshall—map by Maria DiLeo; © Mitsuaki Iwago/Minden Pictures; Simon Fraser/Science Photo Library/Science Source/Photo Researchers. **14, 15:** CORBIS/Rick Doyle—art by Maria DiLeo; CORBIS/Richard Hamilton Smith—illustration by Oliver Rennert, © Weldon Owen Pty. Ltd.; © David Doubilet; © 1998 Jim Brandenburg/Minden Pictures—© Flip Nicklin/Minden Pictures—CORBIS/Tony Arruza—© Frans Lanting/Minden Pictures—art by Stephen R. Wagner. **16:** David Doubilet/National Geographic Society (NGS) Image Collection—© David Doubilet—Mark Wexler/Woodfin Camp & Associates. **17:** Australian Picture Library/John Carnemolla; Annie Griffiths Belt/NGS Image Collection—Peter Parks/Oxford Scientific Films, Long Hanborough, Oxfordshire, England—© 1997 Tim Fitzharris/Minden Pictures; CORBIS/Karl Weatherly. **18:** Art by Maria DiLeo—CORBIS/Owen Franken; Scott Bolden/NGS Image Collection. **19:** Art by Maria DiLeo; © Anthony Bannister/NHPA, Ardingly, Sussex, England—© John Shaw/NHPA, Ardingly, Sussex, England—George Silk, *Life* Magazine, © Time Inc.; CORBIS/Yann Arthus-Bertrand. **20:** Map by John Drummond (2)—James Sugar/Black Star; The G. R. "Dick" Roberts Photo Library. **21:** Tony Arruza/Bruce Coleman Inc., New York; art by John Drummond—© David Doubilet; © 1998 James H. Pete Carmichael Jr.—© 1999 Norbert Wu/www.norbertwu.com. **22:** CORBIS/Mack Gibson; CORBIS/Wolfgang Kaehler. **23:** NASA/Stock Market; NASA/JPL, photo nos. PIA01085 & PIA01461—© American Red Cross/Photo Researchers—© Angus Johnston/Still Moving Picture Company, Edinburgh, Scotland—map by John Drummond; © Gary Bell/Oceanwide Images. **24:** Judith L. Connor; © 1999 Norbert Wu/www.norbertwu.com—© 1993 Andrew J. Martinez—© 1999 Norbert Wu/www.norbertwu.com (3). **25:** Jane Burton/Bruce Coleman Collection, Uxbridge, Middlesex, England; © 1990 Fred Bavendam—J. A. L Cooke/Oxford Scientific Films, Long Hanborough, Oxfordshire, England—© 1996 Andrew J. Martinez. **26, 27:** Art by John Drummond—CORBIS/Bojan Brecelj—Mike Pattisall; © Howard Hall/HHP (background); © David Doubilet; © 1999 Norbert Wu/www.norbertwu.com—© Howard Hall/HHP; © 1999 Norbert Wu/www.norbertwu.com; © Howard Hall/HHP. **28:** CORBIS/Douglas P. Wilson/Frank Lane Picture Assoc. (2); © 1999 Norbert Wu/www.norbertwu.com—CORBIS/Lester V. Bergman. **29:** © Roger Steene (2)—provided by the SeaWiFs Project, NASA/Goddard Space Flight Center and ORBIMAGE. **30:** © Roger Steene (2)—Peter Parks/Oxford Scientific Films, Long Hanborough, Oxfordshire, England; CORBIS/Kevin Schafer. **31:** Ferran Garcia-Pichel/Max Planck Institute for Marine Microbiology, Bremen, Germany; © 1998 E. Widder, Harbor Branch Oceanographic Institution Inc.—Klaus Heumann—© Vieques Conservation and Historic Trust—Frank Borges Llosa@frankly.com. **32:** © Roger Steene—© 1999 Norbert Wu/www.norbertwu.com; illustration by Mike Gorman, © 1991 Weldon Owen Pty. Ltd. **33:** Art by Wood Ronsaville Harlin, Inc. **34:** © Roger Steene. **35:** © 1976 Tom McHugh/Steinhart Aquarium/Photo Researchers—© Roger Steene (2)—© Steven K. Webster—© Mark Spencer/Auscape; art by Will Nelson (3)—art by

Daniel Moyer (4). **36, 37:** © Richard T. Nowitz/Photo Researchers—© Robert Moss Photography, Alexandria, Va.; Nancy Sefton/Planet Earth Pictures, London (background); © Becca Saunders/Auscape—© Fred Bavendam/Minden Pictures; art by Maria DiLeo; © Fred Bavendam/Minden Pictures—from *Biology: The Dynamics of Life,* Glencoe/McGraw-Hill, 2000, Westerville, Ohio. **38:** © Herwarth Voigtmann/Planet Earth Pictures, London—© Al Giddings/Al Giddings Images, Inc.—from *Biology: The Dynamics of Life,* Glencoe/McGraw-Hill, 2000, Westerville, Ohio; © Andrew J. Martinez. **39:** © David Doubilet—art by Kenneth Gosner—© Lennart Nilsson/Albert Bonniers Förlag AB, Stockholm, Sweden (2); © David Doubilet—Dr. P. J. Fenner—© 1987 Fred Bavendam—© 1999 Norbert Wu/www.norbertwu.com. **40:** Becca Saunders/Auscape—Fred Hoogervorst/Foto Natura—Dr. Frieder Sauer/Bruce Coleman Collection, Uxbridge, Middlesex, England; art by Daniel Moyer; © Stan Elems/Visuals Unlimited. **41:** Peter Parks/Oxford Scientific Films, Long Hanborough, Oxfordshire, England (2); Kathie Atkinson/Auscape—© 1999 Norbert Wu/www.norbertwu.com—Becca Saunders/Auscape—© Fred Bavendam. **42:** © Franklin J. Viola—art by Stephen R. Wagner; © 1999 James H. Pete Carmichael. **43:** Art by Jeff McKay; © A. Kerstitch/Visuals Unlimited; Victor Boswell/NGS Image Collection—illustration from *Understanding Science and Nature: Underwater World,* Gakken Co. Ltd., 1990, Tokyo; © Roger Steene—Alex Kerstitch/Planet Earth Pictures, London. **44:** Art by Jeff McKay—© 1999 Norbert Wu/www.norbertwu.com—Becca Saunders/Auscape; illustration by Simone End, © Weldon Owen Pty. Ltd. **45:** © 1988 Fred Bavendam—© F. Stuart Westmoreland/Mo Young Productions—K. Ross/JACANA, Paris—illustration from *Understanding Science and Nature: Underwater World,* Gakken Co. Ltd., 1990, Tokyo; © David Doubilet—© Douglas Faulkner/National Audubon Society Collection/Photo Researchers—© 1999 James H. Pete Carmichael. **46:** © Roger Steene (5); © Rudie H. Kuiter/Oxford Scientific Films, Long Hanborough, Oxfordshire, England—© 1999 Norbert Wu/www.norbertwu.com—© David J. Wroebel. **47:** The Cousteau Society (3); © Rudie Kuiter/Oxford Scientific Films, Long Hanborough, Oxfordshire, England—notebook photo by Mike Pattisall; © 1988 Fred Bavendam (inset). **48:** © David Doubilet—Dorling Kindersley Ltd., London (4). **49:** © Roger Steene—© 1998 Ned DeLoach—© Howard Hall/HHP; David Maitland/Planet Earth Pictures, London—© 1999 Norbert Wu/www.norbertwu.com—© Fred Bavendam. **50:** Fred Winner/JACANA, Paris—© R. A. Preston-Mafham/Premaphotos Wildlife, Bodmin, Cornwall, England; © 1991 Fred Bavendam—Sophie De Wilde/JACANA, Paris—© X. Desmier/Rapho. **51:** © Fredrik Ehrenstrom/Oxford Scientific Films, Long Hanborough, Oxfordshire, England—illustration from *Understanding Science and Nature: Underwater World,* Gakken Co. Ltd., 1990, Tokyo; © Andrew J. Martinez—© Jeanne White/National Audubon Society Collection/Photo Researchers—© Mark Spencer/Auscape—© Steven K. Webster. **52, 53:** © 1999 Norbert Wu/www.norbertwu.com; © Becca Saunders/Auscape (inset); art by Jeff McKay—from *Biology: The Dynamics of Life,* Glencoe/McGraw-Hill, 2000, Westerville, Ohio; © 1999 Norbert Wu/www.norbertwu.com; Heather Angel, Biofotos, Farnham, Surrey, England; Anthony Bannister. **54:** © 1991 Fred Bavendam—© A. Kerstitch/Visuals Unlimited; © 1998 Harbor Branch Oceanographic Institution, Fort Pierce, Fla. **55:** © Andrew J. Martinez—© 1998 Harbor Branch Oceanographic Institution, Fort Pierce, Fla.; © Benelux Press/Stock Market. **56:** Art by Rod Ruth, © 1979 Time-Life Books, Inc., courtesy National Marine Fisheries Service, Chicago (2)—art by Stephen R. Wagner. **57:** Steinhart Aquarium/Tom McHugh/Photo Researchers—© Tom McHugh/Photo Researchers—© Roger Steene—© J. W. Mowbray/National Audubon Society Collection/Photo Researchers; © Tom McHugh/Photo Researchers—© Fred Bavendam—art by Linda Nye. **58:** Max Gibbs/Oxford Scientific Films, Long Hanborough, Oxfordshire, England—© 1992 Charles V. Angelo/National Audubon Society Collection/Photo Researchers; illustration from *Understanding Science and Nature: Underwater World,* Gakken Co. Ltd., 1990, Tokyo. **59:** © David Doubilet; © Doug Perrine/Innerspace Visions; © 1999 Norbert Wu/www.norbertwu.com; © 1994 Brandon Cole/Mo Yung Productions—© 1999 Norbert Wu/www.norbertwu.com; © Mike Johnson. **60, 61:** © David Doubilet—art by Alicia Freile; © Howard Hall/HHP; © Roger Steene; Peter Scoones/Planet Earth Pictures, London; art by Rod Ruth, © 1979 Time-Life Books, Inc., courtesy Marine Fisheries Service, Chicago (4)—CORBIS/Henry Diltz. **62:** Art by

Stephen R. Wagner—illustration from *Understanding Science and Nature: Underwater World,* Gakken Co. Ltd., Tokyo, 1990—© Kevin Deacon/Auscape; © Howard Hall/HHP—illustration from *Understanding Science and Nature: Underwater World,* Gakken Co. Ltd., 1990, Tokyo. **63:** © Roger Steene—© Andrew J. Martinez; © Chris Huxley/Planet Earth Pictures, London—illustration from *Understanding Science and Nature: Underwater World,* Gakken Co. Ltd., 1990, Tokyo. **64:** © 1991 Fred Bavendam—© James H. Pete Carmichael; © Mark Conlin/Mo Yung Productions. **65:** © Rudie H. Kuiter/Oxford Scientific Films, Long Hanborough, Oxfordshire, England; © David Doubilet—© Roger Steene (2). **66:** © 1999 Norbert Wu/www.norbertwu.com—© Tom McHugh/Photo Researchers (2); © Roger Steene. **67:** © Tom McHugh/Photo Researchers—© 1999 Norbert Wu/www.norbertwu.com—© Fred Bavendam—© Roger Steene; © Jeffrey L. Rotman (2)—© 1999 Norbert Wu/www.norbertwu.com—© Roger Steene—© Andrew J. Martinez/National Audubon Society Collection/Photo Researchers. **68:** © Kevin Deacon/Auscape—© David Doubilet. **69:** Illustration by David Kirshner, from *Discoveries: Dangerous Animals,* © Weldon Owen Pty. Ltd.—© Jeffrey L. Rotman (2)—illustration by Will Nelson; © Jeffrey L. Rotman—© Howard Hall/HHP—© David Doubilet—Doug Perrine/Planet Earth Pictures, London—© David Doubilet. **70:** Photo by Nikolas Konstantinou, © David Doubilet—notebook photo by Mike Pattisall; © David Doubilet (2). **71:** © David Doubilet—© Jeffrey L. Rotman—© Tom McHugh/Photo Researchers; © 1999 Norbert Wu/www.norbertwu.com—© Mark Spencer/Auscape—© 1995 Fred Bavendam—© David Doubilet. **72:** © David Doubilet—© Jeffrey L. Rotman—© Andrew J. Martinez; Howard Hall/HHP. **73:** © Mark Spencer/Auscape—Australian Picture Library/Pacific Stock; © Becca Saunders/Auscape. **74:** Map by John Drummond; Fred Bruemmer/Bruce Coleman Collection, Uxbridge, Middlesex, England—art by Will Nelson (2); © Mitsuaki Iwago/Minden Pictures. **75:** © Andrew G. Wood/Photo Researchers—art by John Drummond—Ron & Valerie Taylor/Ardea Ltd., London; Dorling Kindersley Ltd., London—illustration by David Kirshner from *Encyclopedia of Animals,* © Weldon Owen Pty. Ltd.—CORBIS Sygma/Paulo Fridman. **76:** Art by John Drummond—art by Will Nelson (2); © 1999 Carl Roessler. **77:** Art by Carol Schwartz—art by Jeff McKay; art by Jeff McKay—Howard Hall/HHP (2). **78:** Art by Will Nelson; © Jeff Foott (3). **79:** © 1995 Sanford/Agliolo/Stock Market—© 1999 Monterey Bay Aquarium/photo by Rick Browne; Royal British Columbia Museum, no. CPN 11613. **80:** © 1994 ZEFA-Wisniewski/Stock Market—art by Will Nelson. **81:** © Leonard Lee Rue III/Photo Researchers—© Jeff Vanuga—© Eric & David Hoskins/Photo Researchers; © Graham Robertson/Auscape—art by Jeff McKay—Charles W. Fowler/National Marine Mammal Lab, Seattle. **82, 83:** © Douglas Faulkner/Photo Researchers; © Doug Perrine/Auscape; © Douglas Faulkner/Photo Researchers (inset)—The Granger Collection, New York; art by Wood Ronsaville Harlin, Inc.; © Ben & Lynn Cropp/Auscape. **84, 85:** © Michio Hoshino/Minden Pictures; © 1994 Peter Howorth/Mo Yung Productions (inset); art by Heather Lovett (9); art by Robin DeWitt & Pat Grush—Mike Golding, © Weldon Owen Publishing; David Kirshner, © Weldon Owen Publishing; Joyce Photographics/Photo Researchers. **86:** Howard Hall/HHP—art by Will Nelson (2); teeth by Maria DiLeo (2). **87:** Bill Curtsinger/NGS Image Collection—art by Frederico Casteluccio/NGS Image Collection; Naval Command Control and Ocean Surveillance Center, Research Development Test and Evaluation Division, San Diego, Calif. **88:** © Jeff Foott/Auscape—© 1995 James Watt/Mo Yung Productions—CORBIS/Stuart Westmoreland; CORBIS/Joe McDonald; art by John Drummond. **89:** © 1994 Ron Sanford/Stock Market—© François Gohier/Photo Researchers (2); art by Jeff McKay—Christopher Klein/NGS Image Collection. **90:** The Granger Collection, New York—courtesy Mariners Museum, Newport News, Va.—© Flip Nicklin/Minden Pictures; © Tom McHugh/Photo Researchers; Culver Pictures Inc. **91:** © Shane Moore/Animals Animals—© 1997 Amos Nachoum/Stock Market; © Iain Kerr/Whale Conservation Institute—Collection of the Museum of Anthropology, photo by Bill McLennan, cat. no. A63116. **92, 93:** © Kathie Atkinson/Auscape; illustration by Jon Gittoes/Mike Gorman/Oliver Rennert from *Illustrated Library of the Earth: Oceans,* © Weldon Owen Pty. Ltd.; © 1994 Brandon Cole/Mo Yung Productions (inset)—art by Stephen R. Wagner; © Roger Steene—© Graham Robertson/Auscape—James H. Pete Carmichael—© D. Parer & E. Parer-Cook/

Auscape—© 1995 James H. Pete Carmichael. **94, 95:** © Franklin J. Viola—art by Stephen R. Wagner; © Mark Spencer/Auscape; © Roger Steene—art by Stephen R. Wagner. **96:** Map by John Drummond; © Jeff Ripple. **97:** © Michael Yamashita—map by John Drummond—CORBIS/Yann Arthus-Bertrand; © Roger Steene—© Kevin Deacon/Auscape—© David Doubilet. **98:** Art by Maria DiLeo—© Michele Hall/HHP—© Roger Steene; © Kevin Deacon/Auscape—© David Doubilet. **99:** Art by Maria DiLeo—© Roger Steene (2); © Bob Cranston/Mo Yung Productions—© David Doubilet—Steven K. Webster (2). **100:** Margaret Marchaterre; © 1999 Norbert Wu/www.norbertwu.com—Ken Lucas/Planet Earth Pictures, London—© 1999 Norbert Wu/www.norbertwu.com (2). **101:** © 1999 Norbert Wu/www.norbertwu.com—© 1998 James H. Pete Carmichael—Mark Erdmann—Hans Fricke, Max Planck Institut fur Verhaltensphysiologie; art by Stephen R. Wagner. **102, 103:** Art by Maria DiLeo; illustration by Jon Gittoes/Mike Gorman/Oliver Rennert from *Illustrated Library of the Earth: Oceans,* © Weldon Owen Pty. Ltd. (background); © Woods Hole Oceanographic Institution, photo by Rod Catanach; art by Stephen R. Wagner (2)—© Roger Steene; K. L. Smith Jr.; photo by Herve Chaumeton. **104:** Hans Reinhard, Heiligkreuzsteinach—map by Stephen R. Wagner—© 1997 Time Life Inc.—David Rootes/Planet Earth Pictures, London. **105:** © Flip Nicklin/Minden Pictures—© 1999 Norbert Wu/www.norbertwu.com; Nikolas Konstantinou/National Geographic Society—notebook photo by Mike Pattisall—Nick Caloyianis/NGS Image Collection. **106:** Art By Julek Heller, © 1985 Time-Life Books, Inc.—Blauel/Gnamm-Artothek, Peissenberg, Neue Pinakothek, Munich; Mary Evans Picture Library, London. **107:** Woodcuts from *Mostri, Draghi e Serpenti: . . . di Ulisse Aldrovandi,* edited by Erminio Caprotti, published by Nuove Edizioni Gabriele Mazzota, Milan (insets); art by Bryan Leister (background)—Brian Skerry/NGS Image Collection; Oliver Strewe, © Wave Productions Pty. Ltd. **108, 109:** Art by John Drummond; © Michael Holford, Loughton, Essex, England—© National Maritime Museum, Greenwich, London; from *Report of the Scientific Results of the Voyage of HMS Challenger, During the Years 1873-76, Zoology,* Vol. 1, Her Majesty's Stationery Office, London, 1880; Mary Evans Picture Library, London—from *Report of the Scientific Results of the Voyage of HMS Challenger, During the Years 1873-76, Zoology,* Vol. 1, Her Majesty's Stationery Office, London, 1880 (3). **110:** Mary Evans Picture Library, London; Culver Pictures Inc.—Jean-Loup Charmet, Paris—CORBIS/Bettmann. **111:** OAR/National Undersea Research Program (NURP); CORBIS/Bettmann—David Knusen/NGS Image Collection; © 1999 North Wind Picture Archives. **112:** CORBIS/Bettmann—Bates Littlehales/NGS Image Collection. **113:** Charles Nicklin/Al Giddings Images, Inc.—© 1999 Nuytco Research Ltd., North Vancouver, B.C. (3); notebook photo by Mike Pattisall. **114:** © 1998 Harbor Branch Oceanographic Institution, Fort Pierce, Fla.—art by Jeff McKay; U.S. Navy, Submarine Force Museum, Groton, Conn. **115:** Marine Physical Laboratory of the Scripps Institution of Oceanography, UCSD (2); OAR/National Undersea Research Program (NURP) (2)—Murry Sill for Jules' Undersea Lodge. **116, 117:** Larry Sherer, courtesy the George C. Marshall Foundation, Lexington, Va.; © Ken Marschall; Perry Thorsvik/NGS Image Collection (inset)—© 1986 Woods Hole Oceanographic Institution; © Karen Kasmauski/Woodfin Camp and Associates; © 1999 Don Kincaid—© Pat Clyne—courtesy Peabody Essex Museum, Salem, Mass.—map by John Drummond. **118, 119:** © 1998 Discovery Communications, Inc. (6); CORBIS Sygma/© Stephane Compoint (background).

Text Excerpts:
p. 47: "I Was There!" quote excerpted from *International Wildlife* Magazine, January-February 1995, courtesy National Wildlife Federation.
p. 70: "I Was There!" quote excerpted from *Light in the Sea* by David Doubilet.
p. 105: "I Was There!" quote used with permission from National Geographic Society.
p. 113: "I Was There!" quote excerpted from "The Twilight Zone" by Sylvia Earle, from *Saving the Oceans,* general editor Joseph MacInnis, Key Porter Books, 1992.
p. 116: Robert Ballard quote excerpted from *Exploring the Titanic,* Madison Press Books, Toronto, Ontario.

Glossary of Terms

Algae (**al**-jee) A group of simple plants that have no true roots, stems, or leaves.

Animal (**an**-uh-muhl) A living organism that can move by itself, has organs, and does not make its own food.

Bacteria (bak-**teer**-ee-uh) A class of microscopic one-celled or noncellular organisms.

Baleen (bay-**leen**) Sievelike plates in the mouth of certain types of whales that have no teeth; the plates are used for straining food from water.

Bioluminescence (**bye**-oh-loo-muh-nes-uhnss) The making and giving off of light by a living organism.

Bivalves (**bye**-valvz) Mollusks such as clams, oysters, and mussels that have a two-part hinged shell.

Carnivore (**kar**-nuh-vor) An animal that eats meat.

Cartilage (**kar**-tuh-lij) A tough, flexible tissue that provides support and allows movement.

Chordate (**kor**-dayt) A group of animals that have a notochord during some stage of their development; includes all vertebrates and some marine animals.

Circulatory system (**sur**-kyuh-luh-taw-ree **siss**-tuhm) The organs and structures that move blood to and from all parts of an organism; the system includes heart, blood, and blood vessels.

Cold-blooded (**kold-bluhd**-id) Not having the ability to regulate body temperature. The body temperature of a cold-blooded animal fluctuates with the surrounding air or water.

Colony (plural **colonies**) (**kah**-luh-nee, **kah**-luh-neez) A group of organisms of the same species living or growing together.

Community (kuh-**myoo**-nuh-tee) A group of organisms of different species living in the same area.

Condense (kon-**denss**) Change from a gas or a vapor to a liquid.

Crustacean (kruh-**stay**-shuhn) A group of invertebrate animals, including barnacles, shrimp, crabs, and lobsters, with hard, flexible exterior skeletons and two pairs of antennae.

Dense (**denss**) Crowded together so there is more in a given space; the colder or saltier water is, the more dense it becomes. **Density** (**denss**-i-tee) is a measure of how dense a substance is.

Dorsal (**dor**-suhl) On the back.

Echolocation (ek-oh-loh-**kay**-shun) The process of finding an object by using sound waves, which are reflected off the object and back to the sender.

Ecosystem (**ee**-koh-siss-tuhm) The interactions among living and nonliving things in an area, including soil, water, climate, plants, and animals.

Embryo (**em**-bree-oh) The early stages of development of an organism.

Exoskeleton (**ek**-soh-skel-uh-tuhn) A covering on the outside of an animal that gives it support, such as the shell of a crab.

Extinct (ek-**stingkt**) No longer existing or living.

Fertilize (**fur**-tuh-lyz) Start development of a new individual, with the joining of an egg and a sperm.

Flagellum (plural **flagella**) (fla-**jel**-uhm, fla-**jel**-uh) A long, slender, hairlike structure on certain cells that is the main means of motion for many microorganisms.

Gonad (**goh**-nad) Sexual organs that contain eggs or sperm for reproduction.

Habitat (**hab**-i-tat) The natural environment in which a plant or animal lives.

Inorganic (in-or-**gan**-ik) Made of minerals, not plant or animal matter.

Invertebrate (in-**vur**-tuh-brate) An animal that does not have a backbone, or spinal column.

Kingdom (**king**-dum) The largest category of biological classification. All living and fossil organisms are divided into five kingdoms: Monera (bacteria), Protista (generally single-celled organisms), Fungi, Plantae, and Animalia.

Larva (plural **larvae**) (**lahr**-vuh, **lahr**-vay) The immature stage of an animal that changes form to become an adult.

Mammal (**ma**-muhl) A class of warm-blooded animals that nurse their young.

Marine (**muh**-reen) Of the sea.

Medusa (meh-**doo**-suh) The unattached form of a cnidarian such as a free-swimming jelly-fish.

Migration (mye-**gray**-shuhn) A regular seasonal movement of animals to a location with more favorable climate or more abundant food or for the purpose of breeding.

Mollusk (**mol**-uhsk) A soft-bodied invertebrate often enclosed in a hard shell; includes clams, snails, and octopuses.

Molting (**mohl**-ting) The process by which some animals periodically shed their outer covering.

Mucus (**myoo**-kuhss) A thick protective liquid made by animals.

Nervous system (nur-vuhss **siss-tuhm**) The organs and structures of a body, including the brain and nerves, that receive and interpret sensations and transmit impulses.

Nutrient (**noo**-tree-uhnt) Microscopic substances that are produced by digested food and are used by cells for energy, growth, and repair.

Olfactory (ohl-**fak**-tor-ee) Relating to the sense of smell.

Organ (**or**-guhn) A group of tissues, such as the heart or stomach, that work together to perform a specific job.

Organic (or-**gan**-ik) Produced from living organisms.

Parasite (**payr**-uh-syte) An organism that lives in or on another (the host) and obtains food from it, usually harming and often killing the host.

Pectoral fins (pek-**tor**-uhl **finz**) Fins on the sides of fish equivalent to the front legs of four-legged animals.

Photosynthesis (foh-toh-**sin**-thuh-siss) The process in which plants with chlorophyll use water, carbon dioxide, and sunlight to produce sugars and starches.

Phylum (plural **phyla**) (**fye**-luhm, **fye**-luh) The most inclusive category of biological classification that exists within a kingdom.

Pigment (**pig**-muhnt) Matter that colors a cell or tissue.

Pinniped (**pin**-uh-ped) The group of animals that includes seals, sea lions, and walruses.

Plankton (**plangk**-tuhn) Tiny plants or animals that float and drift in water. **Phytoplankton** (**fye**-toh-plangk-tuhn) are plants, and **zooplankton** (**zo**-uh-plangk-tuhn) are animals.

Predator (**pred**-uh-tor) An animal that kills and eats other animals.

Pressure (**presh**-uhr) The force of water or air pressing against a surface.

Prey (**pray**) An animal hunted or caught for food.

Protein (**proh**-teen) A class of molecules with a wide variety of types and functions. Protein is the main component of cells.

Regeneration (ree-gen-uh-**ray**-shuhn) The regrowth of a body part with new tissue.

Reptile (**rep**-tyle) A group of cold-blooded animals that have among their common traits skin made of scales or hard plates.

Salinity (suh-**lin**-i-tee) Saltiness; the amount of dissolved salts in water.

Scavenger (**ska**-ven-juhr) Animals that feed on dead animals.

Sea level (**see lev**-uhl) The level of the sea's surface, which is the point from which the height or depth of land is measured.

Secrete (suh-**creet**) To make and give off from the body.

Sediment (**sed**-uh-muhnt) Particles that settle and accumulate in a loose form.

Siphon (**sye**-fuhn) A tubular organ that is found in many mollusks and other invertebrates and is used to draw in or eject fluids.

Skeleton (**skel**-uh-tuhn) The rigid supportive or protective structure of an organism.

Species (**spee**-sees) The most specific category of biological classification. A species includes organisms that are similar and can breed only among themselves.

Spinal cord (**spye**-nuhl **kord**) The bundle of nerves that runs from the brain down through the spine.

Suspension (suh-**spen**-shuhn) The state of a substance when particles are mixed but not dissolved in a liquid.

Swim bladder (**swim blad**-uhr) The air- or oil-filled sac of certain fish that enables them to float and maintain their depth in water.

Symbiosis (sim-bee-**oh**-suhss) Close association between two organisms that often benefits each.

Symmetry (**sim**-uh-tree) An arrangement, on the opposite sides of a line or around a point, in which parts are equal in size, shape, and position. Organisms with **radial symmetry** (**ray**-dee-uhl **sim**-uh-tree) have similar parts arranged around a point. Those with **bilateral symmetry** (bye-**lat**-uhr-uhl **sim**-uh-tree) have similar parts that are arranged on the opposite sides of a line.

Tectonic plates (tek-**tawn**-ik **playts**) Rigid pieces of earth's crust and upper mantle that carry continents and the ocean floor and move slowly over the surface of the planet.

Tide (**tyde**) The rise and fall of the surface of the oceans, seas, and bays caused by the gravitational pull of the moon and, to a lesser degree, the sun.

Tissue (**tish**-oo) A group of cells that are similar in their composition and that perform a specific job.

Trench (**trench**) A deep, long valley on the ocean floor.

Vapor (**vay**-puhr) A mist or mass of tiny drops of water floating in the air; the gas form of a substance, as in water vapor.

Vertebrate (**vur**-tuh-brate) An animal having a backbone or spinal column.

Warm-blooded (**wawrm-bluhd**-id) An animal that maintains a constant body temperature, independent of the surrounding temperature.

Index

124

Index

TIME® LIFE BOOKS

Time-Life Education, Inc. is a division of Time Life Inc.

TIME LIFE INC.

PRESIDENT and CEO: George Artandi
CHIEF OPERATING OFFICER: Mary Davis Holt

TIME-LIFE EDUCATION, INC.
PRESIDENT: Mary Davis Holt
MANAGING EDITOR: Mary J. Wright

Time-Life Student Library
OCEAN LIFE

SERIES EDITOR: Jean Burke Crawford

Associate Editor/Research and Writing: Mary Saxton
Series Picture Associate: Lisa Moss
Editorial Assistant: Maria Washington
Picture Coordinator: Daryl Beard

Designed by: Jeff McKay and Phillip Unetic, 3r1 Group

Special Contributors: Susan S. Blair, Scarlet Cheng, Jim Lynch, Susan McGrath, Susan Perry, Terrell Smith, Barry Wolverton (text); Patti Cass, Connie Contreras (picture research); Barbara Klein (index).
Senior Copyeditor: Judith Klein
Correspondents: Maria Vincenza Aloisi (Paris), Christine Hinze (London), Angelika Lemmer (Bonn), Christina Lieberman (New York).

Senior Vice President and Publisher: Rosalyn McPherson Perkins
Sales Director and Associate Publisher: Cheryl Crowell
Vice President of Marketing and Promotion: David Singleton
Director of Book Production: Patricia Pascale
Director of Publishing Technology: Betsi McGrath
Director of Photography and Research: John Conrad Weiser
Production Manager: Vanessa Hunnibell
Director of Quality Assurance: James King
Chief Librarian: Louise D. Forstall

Consultants: Peter Petraitis, Ph.D., is a marine ecologist at the University of Pennsylvania, where he is a professor of biology and director of the Institute for Environmental Studies. He has been fascinated with the sea and marine life since his first encounter with a live lobster at the age of three. His research focuses on the ecology of rocky intertidal shores and the biology of marine invertebrates. He has studied the ecology of marine invertebrates on both coasts of the United States and in Australia.

Lisa Lyle Wu teaches biology, marine biology, and advanced-placement biology at Thomas Jefferson High School for Science and Technology in Fairfax County, Virginia, and has taught biology since 1978. She has extensive experience in development and execution of curricula for biology and oceanography. Her research has included the study of sea scallops, sea urchins, squirrelfish, dolphins and whales, and coral reef and tropical island ecology. She is also an education specialist for the Smithsonian Institution's National Museum of Natural History and has developed curriculum materials for the Discovery Channel and the National Audubon Society.

First printing. Printed in U.S.A.
School and library distribution by Time-Life Education, P.O. Box 85026, Richmond, Virginia 23285-5026.
Telephone: 1-800-449-2010
Internet: www.timelifeedu.com

TIME-LIFE is a trademark of Time Warner Inc. and affiliated companies.

Library of Congress Cataloging-in-Publication Data
Ocean life
    p. cm. — (Time-Life student library)
    Includes index.
    Summary: Examines a variety of ocean life, including marine plants, fish, reptiles, and mammals, and discusses their habitats, reproduction, and defenses.
    ISBN 0-7835-1357-7
    1. Marine biology Juvenile literature. [1. Marine animals. 2. Marine biology.]
I. Time-Life Books. II. Series.
QH91.16.O3 1999
578.77—dc21   99-29720
                                                                                  CIP

OTHER PUBLICATIONS

TIME-LIFE KIDS	SCIENCE/NATURE
Library of First Questions and Answers	Voyage Through the Universe
A Child's First Library of Learning	DO IT YOURSELF
I Love Math	Total Golf
Nature Company Discoveries	How to Fix It
Understanding Science & Nature	The Time-Life Complete Gardener
	Home Repair and Improvement
HISTORY	The Art of Woodworking
Our American Century	
World War II	
What Life Was Like	COOKING
The American Story	Weight Watchers® Smart Choice
Voices of the Civil War	Recipe Collection
The American Indians	Great Taste-Low Fat
Lost Civilizations	Williams-Sonoma Kitchen Library
Mysteries of the Unknown	
Time Frame	
The Civil War	
Cultural Atlas	

For information on and a full description of any of the Time-Life Books series listed above, please call 1-800-621-7026 or write:

Reader Information
Time-Life Customer Service
P.O. Box C-32068
Richmond, Virginia 23261-2068